Ewald R. Weibel is Professor of Anatomy, Emeritus, University of Berne. He is the author of *The Pathway for Oxygen* (Harvard).

Symmorphosis

Symmorphosis

On Form and Function in Shaping Life

Ewald R. Weibel

Harvard University Press
Cambridge, Massachusetts
London, England · 2000

Copyright © 2000 by the President and Fellows of Harvard College
All rights reserved
Printed in the United States of America
Book design by Dean Bornstein

Library of Congress Cataloging-in-Publication Data

Weibel, Ewald R.
   Symmorphosis : on form and function in shaping life / Ewald R. Weibel
     p. cm.
   Includes index.
   ISBN 0-674-00068-4 (alk. paper)
   1. Symmorphosis. I. Title.
QH491.W447 2000
571.3′1—dc21   99-4347

*To the memory of Charles Richard Taylor (1939–1995)
Best partner—best friend*

Contents

	Preface	xi
1.	**Form and Function**	1
	The Relation of Form and Function	5
	Adaptation of Function as a Design Principle	8
	Integration of Function as a Design Principle	11
	Economy as a Design Principle	12
	The Principle of Symmorphosis	18
2.	**Cells and Tissues: Oxidative Metabolism in Muscle**	24
	Energy Supply and Mitochondria	29
	Is Mitochondrial Structure Matched to the Demand for Oxidative Energy?	35
	Testing for a Quantitative Match of Form and Function in Muscle Mitochondria	41
	Is $\dot{V}_{O_2}$max Related to V(mi) in Exercising Muscle Cells?	45
	Natural Variation in Energy Demand and Mitochondria	47
	Are Muscle Capillaries Adjusted to Mitochondrial Oxygen Needs?	52
	Symmorphosis in the O_2 Pathway in Muscle	56
3.	**Muscle: Supplying Fuel and Oxygen to Mitochondria**	60
	Differences between Oxygen and Fuel Supply	61
	Variations in Fuel Supply to Mitochondria in Working Muscle Cells	66
	Partitioning of Fuel Consumption between Glucose and Fatty Acids	68
	Estimating the Capillary Supply of Substrates	69
	Revising the Model for Capillary Oxygen and Substrate Supply	71

Fuel Supply from Capillaries versus Intracellular Stores ... 74
Conclusions on Form and Function in Muscle Cells and Tissue ... 77

4. Organ Design: Building the Lung as a Gas Exchanger ... 80
Modeling Gas Exchange in the Lung ... 85
A Large Surface and a Thin Barrier Determine the Gas Exchange Capacity of the Lung ... 90
The Diffusing Capacity of the Human Lung ... 95
How Much Lung Diffusing Capacity Do We Really Need? ... 101
The Gas Exchanger of the Athletic Pronghorn ... 105
The Effect of Reducing the Gas Exchanger ... 107
Conclusion ... 109

5. Problems with Lung Design: Keeping the Surface Large and the Barrier Thin ... 110
A Fiber Continuum Supports Parenchymal Structures ... 111
Controlled Surface Tension Determines Parenchymal Mechanics ... 116
Keeping the Barrier Dry and Thin ... 124
Conclusion ... 130

6. Airways and Blood Vessels: Ventilating and Perfusing a Large Surface ... 131
Morphogenesis of Airways, Vessels, and Gas Exchanger ... 132
Designing the Airway Tree for Efficient Ventilation ... 135
Are Airways Designed as Fractal Trees? ... 145
Conclusion ... 149

7. The Pathway for Oxygen: From Lung to Mitochondria ... 150
Testing the Hypothesis of Symmorphosis ... 152
The Strategy: Exploiting Comparative Physiology ... 154
The Model and Predictions ... 158
Testing the Respiratory System for Symmorphosis ... 162
Does Symmorphosis Prevail in the Respiratory System? ... 178

8.	**Adding Complexity in Form and Function: The Combined Pathways for Oxygen and Fuels**	**181**
	Strategies for Oxygen and Fuel Supply	183
	Design of the Fuel Supply Pathway	184
	Design of Nutrient Uptake Systems	185
	The Substrate Pathways for Fueling Muscle Work	195
	The Test of Symmorphosis	197
	On Symmorphosis in Complex Pathways	206
9.	**Symmorphosis in Form and Function: Concepts, Facts, and Open Questions**	**210**
	How to Perform a Test of Symmorphosis: Future Prospects	213
	Conclusions	218
	References and Further Reading	231
	Credits	255
	Index	259

Preface

This book is the result of often heated discussions that were an important part of twenty years of fruitful collaboration with C. Richard Taylor, my partner and friend at the Museum of Comparative Zoology of Harvard University. These discussions, in a way, culminated in the preparation of the John M. Prather Lectures, which I was asked to present in the Biology Department of Harvard University in May 1995. Our conversations were put to an end a few months later by the untimely death of Dick Taylor on September 10, 1995. This book is therefore dedicated to the memory of my good friend and partner with whom I had the privilege of sharing much excitement in the pursuit of our research into the quantitative relations between structure and function.

Our partnership began in May of 1975. We had worked in totally different fields: Dick Taylor on the efficiency of animal locomotion, and I on the role of structural design in making a well-functioning lung. By chance I had read one of Taylor's papers on the energetics of running animals, and this seemed to suggest a solution to one of the paradoxical problems I had encountered in my research on the lung. We met over lunch at the Harvard Faculty Club and decided to launch a joint expedition to Kenya to study the relationship between muscle energetics and the lung as gas exchanger in large wild mammals. We went to Africa in 1977 with a simple question and returned with not only a truckful of materials but ideas that would occupy us for the rest of Dick Taylor's life. The simple question was whether the lung was adjusted to the maximal rate of oxygen consumption in running animals of different body size. But it soon became apparent that this was not enough of a challenge. Sitting in the shade of a tree at the University of Nairobi we decided to expand the study to look at the design and function of the entire respiratory system, from muscle mitochondria to the lung, what we would later call the pathway for oxygen. This was the beginning of a long story that eventually addressed broad questions of integrative physiology, from the molecular structure of mitochondria in muscle cells to the entire system for energy supply, including respiration and fuel supply and combining physiological, biochemical and morphometric methods.

When, in 1980, all the material from the African study was evaluated, the entire team met in Berne to write up the functional and structural results into a series of papers. This is when the principle of symmorphosis was formulated: it postulates that the quantity of structure incorporated into an animal's functional system is matched to what is needed: enough but not too much.

The work continued over the years. Symmorphosis became a guiding principle in our quest to understand the complex relations between the design of the structures that support functional processes. In my Prather Lectures I pulled together all the work we had done over two decades with a large number of collaborators, and tried to put it into the broader perspective of the relation of form and function in animal design. The present book is a tribute to the guiding spirit of my friend Dick Taylor. I take this opportunity to express my deep gratitude, to him posthumously and to the many collaborators that have joined us in this exciting endeavor over the years,* mainly those who have collaborated for many years and have taken on major responsibilities, not only my closest collaborator, Hans Hoppeler, but also Hans Bachofen, Peter Gehr, Stan L. Lindstedt, Odile Mathieu, Susan Kayar, Kevin E. Conley, Richard H. Karas, and James H. Jones.

I also wish to thank those who have helped me in preparing this book. Alfred W. Crompton from the Museum of Comparative Zoology, Dick Taylor's close friend, has been an important source of encouragement over the years. My secretary, Elsbeth Hanger, and Karl Babl, Eva

---

*In addition to those named below, these include R. McNeill Alexander, Alex Ammann, Robert B. Armstrong, Marianne Bachofen, J. Eduardo P. W. Bicudo, Rudolf Billeter, Gérard Brichon, Peter H. Burri, Helgard Claassen, Minerva Constantinopol, Luis Cruz-Orive, Fabienne Doffey, Norman C. Heglund, Hans Howald, Connie Hsia, Rudolf Krauer, Vaughn Langman, Arne Lindholm, Kim E. Longworth, Geoffrey M. O. Maloiy, Alfred E. Müller, Deter K. Mwangi, Eviatar Nevo, George Ordway, R. Blake Reeves, Thomas Roberts, Kai Rösler, Samuel Schürch, Annemarie Geelhaar Schweingruber, Klaus Schwerzmann, Howard Seeherman, Vilma Stalder-Navarro, Ruth Vock, Eva Wagner, Jean-Michel Weber, Hans-Rudolph Widmer, David Xiao-san Wu, and Georges Zwingelstein.

Wagner, and Barbara Krieger offered invaluable assistance. At Harvard University Press, I thank Michael Fisher for his constructive criticism and for aiding in the preparation of the book for publication and Elizabeth Gilbert for her valuable editorial work. Special thanks go to my reviewers for their most useful comments, and to Hans Hoppeler for his excellent critique of the manuscript.

Our work has been generously supported over the years by the Swiss National Science Foundation and, in the United States, by the National Institutes of Health and the National Science Foundation. This is gratefully acknowledged; I also acknowledge with gratitude the support we have received from our universities, Harvard and Berne, and from the M. E. Müller Foundation.

Berne, July 1999

Symmorphosis

1: Form and Function

> *In general no forms exist save such as are in conformity with physical and mathematical laws.*
>
> D'Arcy W. Thompson, *On Growth and Form*

Very few books leave you with a lasting impression that determines your philosophy, the central goals that you pursue. For me such a book was D'Arcy Wentworth Thompson's *On Growth and Form,* where he shows how structures both beautiful and useful develop during the morphogenesis—the formation of new structure—of organisms, and how growth leads to proportions between the parts that make sense. I read this book in 1959 when I was trying to find my way in science, and I have kept it in my bookshelf at an easy reach from my desk ever since.

D'Arcy Thompson used mathematical models to explain how and why the forms of living creatures came about and what the effects of size and proportions were. And he used physical laws to understand the logic of certain forms. What I wish to do in this book is to pursue similar lines of argument but to stress what form means for functional performance. My thesis is that the structural design of organisms is closely adjusted to functional needs, and that good function is ensured by good "engineering" in all parts of complex organisms. Of course "engineering" should not be taken literally. As I discuss later, what I call here "engineering" is, in fact, the result of evolution, of mutations in the genetic blueprint that lead to variations in structure and function. These changes appear to be the result of engineering if they are well matched.

When we look at nature we find impressive examples of form closely related to function in biological organisms. Consider the pronghorn, an antelope-like animal of the Rocky Mountains (Figure 1.1), the fastest endurance runner among mammals, capable of running across the vast prairies of Wyoming at speeds of up to 60 kilometers an hour and maintaining this speed for nearly an hour without pausing. When we

2 · · · Symmorphosis

Figure 1.1 Pronghorn of the Rocky Mountains *(Antilocapra americana)*.

look at the pronghorn we notice the extraordinarily slim long legs, adapted for long strides. We also notice the large muscle mass of the upper hind legs that allows the animal to accelerate with a powerful thrust. Quite clearly, the form of the pronghorn is beautifully matched to the function of fast running, which it needs to escape predators.

Long slender legs are a trademark of high performance runners, racehorses for example, but also humans. If you watch an athletic competition, you will see that the champion long-distance runners are very slim and long-legged, with a reasonable but not excessive muscle mass in their legs; this allows them to propel their body at high speed and maintain that speed for a long time. In contrast, wrestlers are stocky and heavy-set, as are shot-putters, who have developed a powerful muscle mass in their arms and legs to allow them to heave the steel ball with great force. This type of athletic build is also found in the animal kingdom; for example, in a bull with its stocky legs and thick neck.

These are gross, visible differences in body form related to different patterns of functional performance, but there are also more subtle, hidden differences. To cite just one example here, the bones of the serving arm of a champion tennis player are thicker and thus stronger than those in the other arm so as to compensate for the strain on the bone during the service stroke. Many more examples of such differences will appear in later discussions.

Consideration of the relation between form and function in biological organisms has an aesthetic aspect. We may ask whether whatever beauty we find in the form of organisms is rooted in their inner structural and functional harmony, since the outer form of the body is necessarily the result of inner design. In doing so we would be following the line of thought that D'Arcy Thompson pioneered when he strove to explain, from the first principles of physics and mathematics, the growth and development of biological forms—ranging from foraminiferous shells to blood vessels and bones. He showed that morphogenesis leads to organisms built to suit the functions they perform, to forms well adapted to the prevailing life conditions, as a fish's body is adapted for swimming and a bird's wing for flying. He also suggested that engineers can learn a lot from the solutions to "engineering" problems found in nature, mainly in the living world. When he discussed the intricacy of the form and construction of bones that allows them to support loads efficiently, he compared bone design with an engineer's feat in building a bridge with minimal material. Indeed, the slender limb bones of the pronghorn could not support its weight, particularly not on ground impact during running, if the structure of the bone were not very precisely engineered, including the gross disposition of bone to make a tubelike shaft of appropriate dimensions and a trabecular support of the joint surface as well as the submicroscopic arrangement of collagen fibers within the bone to give it the required tensile strength.

In this book I will, however, elaborate on the *functional importance of form* in biological organisms. I submit that form is not an end in itself but that it serves a function that advances the higher interests of the organism. Biological form is the result of development and growth on the basis of a genetic blueprint that determines how the parts are built and assembled. This holds for the entire organism as well as for its organs, which are themselves aggregates of tissues and cells. Since the organism is a highly complex aggregate of a vast range of different units that interact with each other in one way or another, it is intuitively plausible that a good balance between the parts is required to ensure adequate function, which is itself the complex result of these interactions. For the pronghorn to run at high speeds for an extended period, it obviously requires more than elegant long legs. To perform well-coordinated and sustained powerful contractions, the muscles depend on the activity of two functional systems of the organism: they must be controlled by the nervous system, from the brain down to the fine nerve fibers that connect to each muscle fiber, and they must be supplied with an adequate amount of energy by what we may call the energy-supply system. This is not a simple matter, because it requires the concerted efforts of the heart, the lung, the gut, the liver, and the blood to supply the muscle cells with the quantities of oxygen and fuels that they need in order to generate the energy required to support muscle work.

In this discussion of the relation of form and function I will use the system for energy supply as a model case mainly because it shows a high degree of complexity, involving several organ systems and reaching from macroscopic structures down to the cellular and molecular level. It is also a system that pervades the entire organism since all cells, whether in the brain, the skin, or muscle, require energy for their work.

My basic thesis is that form—or, as I like to call it, the structural design of the organism and its parts—is tightly adjusted to functional need, not only qualitatively but quantitatively as well. It is the proportions between the design elements that are the key to the functional performance of an organism. The gross examples discussed above show largely quantitative structural differences, for example in the shape and size of the legs, that give the body a capacity for a specific functional performance. We will see that this kind of structural adjustment related to different functions occurs in many instances and at all levels of organization from the cells to organs and organ systems. The question I wish to ask and explore is whether this tight relation between form and function is a basic principle of organismic design, one that reflects "good engineering" of the body. I will deliberately focus on terrestrial mammals, of which the human is a special case. These mammals are all built on the same basic body plan, but they show important variations that we can use to see how design is varied to ensure adequate performance of the organism under different conditions.

But before beginning this discussion we need to clarify a few basic concepts, in particular the difference between form and function, and the role of three basic design principles: adaptation, integration, and economy.

The Relation of Form and Function

Let us now use a case study of central biological importance to see how form and function interact. The case is how the heart achieves and adjusts blood flow to serve the supply of oxygen (O_2) to the muscles as they increase their work, for example in running. Figure 1.2a shows how the human heart acts as a cyclical pump: blood flows into the heart chamber (open arrow) during diastole when the heart muscle relaxes and the chamber widens; when it is filled the muscle contracts (short arrows) in what is called systole and blood is expelled into the aorta (solid arrow), from where it flows through the arteries to the cells of working organs. The total blood flow is the product of the volume of blood ejected with each beat, called the stroke volume, and the heartbeat frequency, that is, the number of beats per minute. In the normal human adult at rest, total blood flow is about 5 liters per minute, and this is achieved with a stroke volume of about 80 milliliters and a heart rate of 60 per minute.

When we begin to run at increasing speed the O_2 needs of the muscles increase up to a maximum that is about ten times higher than the resting O_2 consumption level (Figure 1.2b). This increased O_2 need requires increased blood flow. How does the heart achieve this? We all know that heart rate increases as we run harder, and this is indeed the cause of the increase in blood flow; the heartbeat frequency increases, whereas the stroke volume remains essentially unchanged. At the maximal rate of O_2 consumption by locomotory muscles the heart rate has also reached its maximum (Figure 1.2c).

The first point I want to make with this example is that the performance (blood flow) of an organ (the heart) is determined by functional as well as structural parameters. In this case the stroke volume is a

6 · · · Symmorphosis

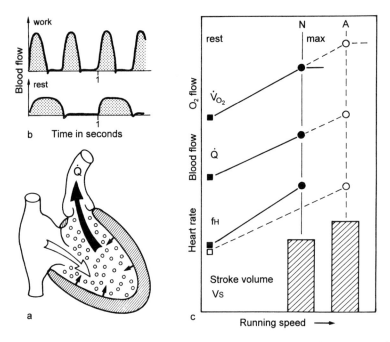

Figure 1.2 Factors determining blood flow in the human. (a) The heart as a cyclical pump. (b) Blood flow ($\dot{Q}$) is pulsatile, with pulses more frequent at work than at rest, but blood flow per beat (area of the curve) is unchanged. (c) The heart rate (f_H) is increased to provide higher blood flow as oxygen consumption ($\dot{V}_{O_2}$) increases, up to maximal work. Athletes (A) increase oxygen supply to their muscles through an increase in the size of the heart ventricle, allowing for a larger stroke volume (V_s), but their maximal heart rate is the same as that of nonathletes (N).

structural parameter because it is directly related to the size of the heart chamber, which fills in diastole and empties nearly completely during systole. Heartbeat frequency is a functional or physiological variable that is capable of being varied rapidly over a wide range according to need.

The case shows that there is a limit to each of these functions: O_2 consumption, blood flow, and heartbeat frequency, and these limits seem to occur at about the same work level. What happens then if an athlete, say a long-distance runner, increases his performance by training? First he will achieve a higher maximal level of O_2 consumption and reach it at a higher running speed, and also his blood flow will be increased according to need (Figure 1.2). But it turns out that his maximal heart rate remains unchanged. What has changed is the stroke volume, which is now large enough to achieve a higher blood flow at the same heartbeat frequency. The larger stroke volume is the result of the increase in heart size that was achieved by training for higher performance. This adjustment of structure to altered function is a slow process that typically takes weeks to months—or even years—to be completed because it involves morphogenesis, in this case the formation of additional heart muscle mass to maintain the thickness of the pump wall while increasing the pump volume.

This example shows the difference between functional and structural design parameters. *Functional variables,* such as heart rate, allow rapid adjustment of performance to actual needs, adjustments that occur in time scales of milliseconds to seconds and minutes. They, in fact, allow the economic operation of the system, in this case by keeping blood flow as low as possible and only increasing it when the need arises. In contrast, *structural design parameters* are "fixed" values with respect to the operation of the system. Together with the maximal rate of the associated functional variable(s), they set the functional capacity of the system, here maximal total blood flow. The body can, as we have seen, adjust such a capacity to higher levels if the functional need occurs, and does this by enlarging the design parameter—heart size to enlarge stroke volume in our example—but such adjustment occurs very, very slowly. It is particularly important to note that this ability to enlarge structures entails economy: it allows structures to be kept as small as is compatible with the expected maximal function until the need for enlargement arises. Indeed, the heart gradually shrinks back to its normal size once the athlete stops his high-performance training.

Is this distinction between functional variables and design/structural parameters of broad enough significance to be of value for discussion of the relation of form and function? I believe so. We can find many cases where the functional capacity is set by a maximal functional rate and a design parameter. To mention only one example from biochemistry: enzyme systems operate at a variable rate according to the needs of the cell up to a maximal or limiting rate, V_{max}, but their total maximal performance is determined by V_{max} and the concentration or content of the enzyme in the cell system. The number of enzyme units built into a membrane or an organelle is clearly a design parameter whose upregulation requires the synthesis of new enzyme and is therefore a comparatively slow process.

Adaptation of Function as a Design Principle

Animals come in all sizes, and they have adapted to fill all sorts of niches in our varied world. The largest animals are mammals such as elephants and whales weighing several tons, but mammals range in size from these giants, in a more or less continuous sequence, down to the very smallest such as the bumble bee bat and the Etruscan shrew, which weighs no more than 2.5 grams as an adult. In contrast, the largest insects weigh only about 30 grams and the great majority of them are very small, the fruit fly *Drosophila* weighing only about 1 milligram. The reason that there is no mammal as small as a fly, or no insect as large as a rabbit, may well have to do with limits to the adaptability of the different design elements of the two types of organisms, for example the simple system of air-filled tracheoles that supplies oxygen in insects compared with the complex blood-based oxygen-transport system in mammals.

One of the important adaptations of animals is to their food sources. Shrews are insectivores; they feed on animals smaller than themselves—though sometimes not much smaller—that have a high caloric value, which is essential for shrews because they have a very high metabolic rate even at rest, as we shall see. Other mammals are either herbivores or carnivores, or omnivores, and they have gastrointestinal systems adapted to coping with the special conditions of digesting their preferred food. Thus in the course of evolution, many herbivores have developed special mechanisms in the stomach or hindgut that fermentatively break down otherwise indigestible plant fibers such as cellulose, which constitute a large proportion of the plant material they collect by foraging. This adaptation of the design of the gut is of great importance because it allows the animals to select a suitable environmental niche, and specifically to feed on a variety of nutrient materials, while still permitting the *internal* machinery of all mammals to operate with essentially the same raw materials—glucose, fatty acids, and amino acids—and to maintain very similar internal conditions in their tissues.

The variation of animal design we observe in nature is the result of the evolution of species by natural selection as described by Darwin. This long-term process favors the survival of the fittest, the perpetuation of the genes of those animals that are best adapted to the conditions of their ecological niches, such as environment, food availability, or predators. This gave them a reproductive advantage, enabling them to produce more progeny than other animals. I will not discuss the mechanisms of evolution; what is important here is that the structural and functional characteristics of the different species are imprinted in each individual's genome, which contains the entire blueprint for the design of the organism. Evolution results from chance mutations in some parts of the genome that alter some design properties as well as the functions associated with them. A mutation that imparts an advantage will, over time, replace less favorable characteristics, allowing the organism to adapt to specific conditions. This process is what I refer to when I talk about "design" or "engineering"; such terminology should not be taken to imply that evolution proceeded by any means other than mutation by chance and selection of the fittest.

Recent years have brought enormous progress in the deciphering of genomic information as well as in the understanding of how this information affects the design of the organism. One important insight has been that the genome of animals, particularly of mammals, is very conservative—that much of the genomic information is common to all or many species in a phylum, and partly even to animals in different phyla; the eye genes in flies and humans, for instance, have many commonalities, although the design of the eye in the two organisms is very different. The conservative nature of the genome is mostly shown in the elements of internal design at the level of the cells, rather than higher-order structures. The differences between the structure of muscle cells in the human, the frog, and the honey bee are much smaller than the differences between the organisms themselves. And among mammals the basic design of the heart or the lung is the same, as is the design of the locomotor system or the brain. Mammalian species differ mostly by quantitative traits, by often subtle shifts in the proportions. Thus athletic species have a larger muscle mass than more sedentary species of the same body size, a variation resulting from adaptation to different life styles during evolution.

There is, however, a different type of "adaptation" that I must discuss. It is the capacity of organisms to quantitatively adjust design and function in response to altered demand. This is a widespread phenomenon that ranges from the bronzing of the skin on sun exposure to improve its filtering capacity for ultraviolet light to the development of a larger muscle mass in response to extended physical exercise and the concomitant increase in heart size to cope with the higher demand for blood flow. Other examples are the induction of enzymes for specific metabolic tasks and the development of specific antibodies in response to antigen stimulation. In fact, nearly all cellular functions are of this nature: they are normally maintained at low activity and are only induced to increase when the need occurs. This malleability demonstrates the adaptability of organismic design to functional need.

In sum, we have observed three causes for variations in mammalian animal design. First, body size causes differences between species that can be subsumed under the term *allometric variation*. Second, programmed "adaptation" to environment, life style, or expected loads on the organism causes great variety in nearly all size classes, a phenomenon called *adaptive variation*. Finally, the cells and organs of animals are able to adapt to different loads, at least to a certain extent; this ability is described as *induced variation* because it is an epigenetic phenomenon that acts at the level of the phenotype and leaves the genotype unaffected. In contrast, allometric and adaptive variations are basically genotypic phenomena.

Before ending this discussion I should say something about the conditions under which adaptation of the organism and its parts occurs, and by which it is possibly limited. First of all, there are *historical constraints* in that evolution by natural selection may have occurred at a time when conditions were different from those that now prevail, so that we do not know why a specific adaptation was favored. In particular, environmental pressures are unpredictably variable in time and space, and therefore a trait that now appears far from optimum may have been a useful adaptation in the distant past, and vice versa.

Most of the discussion of adaptation in evolution is concerned with *external constraints* such as environmental pressures or resistances to which the organism must respond. But there are also *internal* or *constitutional constraints,* generated by the great inner complexity of the organism with the inner logic of interconnected functions. These internal constraints are due to the highly conservative internal constitution of the body. If the difference between the muscle cells of the frog and of the dog is smaller than the differences between the organisms, then this difference becomes even smaller if we look at mitochondria, at myofibrils, or at metabolic pathways and their enzymes.

External and internal constraints are in a way conflicting constraints among which the organism must seek a reasonable compromise when adjusting its design to varying functional demands. Indeed, the organism is at the intersection between the inner and the outer world and, in natural selection, "what determines the success of an individual is the ability of the internal machinery of the organism's body . . . to cope with the challenges of the environment" (Mayr 1991, p. 87). But the internal machinery must follow its own rules when the different parts are required to work in concert.

Integration of Function as a Design Principle

The organism is made of a very large number of specialized parts that are all dependent on one another; it is a highly "democratic" and "social" structure with no part superior to the others. Even the central nervous system is just as much the servant as it is the manager of the body. The main characteristic of the organism is its high level of complexity, not just in terms of composition but, more important, in terms of the interaction and interrelation between the parts at all levels of organization: from the cells and their organelles to the organs and the organ systems, the organism displays a complexity that is at least as great as that of a large city. This interaction results in what one calls emergent properties of the complex system, those properties that "are not there" but occur through the concerted interaction of the parts. The organism thus becomes a whole that is more than the sum of its parts. One example of such emergent properties, relevant here, is locomotion, which comes about by the action of a complex set of muscles on a skeleton made of bones and joints, and this under the direction of the nervous system and with energy provided by fuels and O_2 supplied from the environment.

The integration of complex functions plays a central role in the relation between form and function in biological organisms. And structural design is a key feature of functional integration at all levels of organization. It is structural design that makes the difference between the liver and a bag full of enzymes—even if the enzymes are there in correct proportions and quantities. Structural design puts each element into its proper place in the hierarchy and assembly of the organism: it organizes organelles and membranes in the cells, it assembles cells into tissues, it combines tissues and cells to make organs with their precise order, and it joins organs to make functional systems and, finally, the whole organism. This entire process is based on architectural and engineering blueprints that are gradually executed during development on the instruction of special sets of genes, such as the homeotic genes. In engineering terms, it is structural design that makes cables and tubes for the conduction of neural messages and the flow of blood, and membranes of all sorts to join compartments at interfaces while keeping them separate. Structural design constructs the framework within which functional processes can take place in an orderly fashion, and thus establishes the key relation between form and function.

Economy as a Design Principle

In engineering, it is important to design a machine for the most economical functioning, and to build it as cheaply as possible. The underlying principle is to be economical with resources both in the construction process and in the operation of the machine. Is it also a principle in biology to build organisms economically, not to waste resources and to achieve the least expensive operation? What I have said about adaptation, regulation, and integration suggests that this may be the case, but we need to look for more objective, quantitative evidence before we can say that economy is a design principle in biology.

Is there, in the first place, evidence that functional performance is regulated to be as economical as possible? Yes, for example, from a case study done by Donald Hoyt and C. Richard Taylor in which they asked why horses change their gait from a walk to a trot to a gallop when they increase their running speed. They studied horses running on a treadmill and found that overall the O_2 consumption as a measure of energy expenditure increased linearly. But at low speeds, when the horses were walking, O_2 consumption increased steeply as they approached 1.5 meters per second, the point at which they would change into a trot. A similar pattern emerged as they approached 4.5 meters per second,

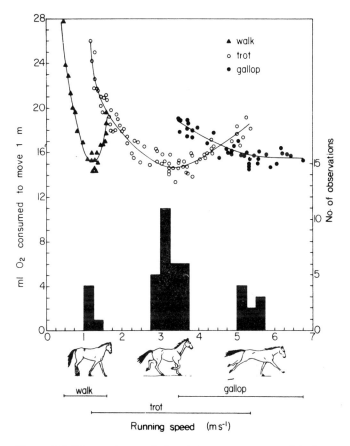

Figure 1.3 Horses change their gait in an attempt to minimize energy expenditure as running speed increases. (From Hoyt and Taylor 1981.)

when they changed into a gallop. These findings suggested that horses change gait in order to work more economically. And indeed, when Hoyt and Taylor looked at their measurements more carefully, they saw that in each gait there was a speed where the energetic cost of moving the body by 1 meter was minimal, with greater costs at slower and faster speeds (Figure 1.3). Most interesting of all, the minimal cost of moving a given distance was the same in all three gaits. Then Hoyt and Taylor observed one of the horses freely running in a field and found that it clearly preferred the gait and the speeds where the cost of locomotion was minimal. The horse never ran for any length of time at a speed where the energy expenditure was higher, a very remarkable finding.

This experiment shows explicitly that functional performance is regulated in animals to accomplish tasks in the most efficient way; it shows that minimization of energy expenditure, or economy, is a driving force in regulating function. Is this also a principle in designing the structure of the organism?

As a first point note that economic design appears to be rooted in the way cells are designed to perform their specific functions. Although all cells contain the entire genome, which defines the complete body plan, only the part relevant for the cell's specific function is expressed. Intestinal epithelial cells, for example, are designed for nutrient uptake and processing, and for that purpose they express the genes for a set of specific enzymes built into their membranes. What is not needed is suppressed; they do not, for example, express enzymes for collagen fiber synthesis, as they are found in fibroblasts, nor the genes for hemoglobin. In contrast, erythrocytes, being designed to transport oxygen in the blood, during their development express mostly the genes for hemoglobin and for a few structural proteins. Eventually, mammalian erythrocytes even get rid of their nuclei, because they are no longer needed once the cells are discharged from the bone marrow into the blood. Even liver cells, these large cells that serve as the main chemical factory of the body, express only a small fraction of the genes contained in their large nuclei.

These few examples should suffice to establish the principle that cells do not burden themselves with baggage they don't need. Each of the cells of the body is specialized for a specific task and this is best served if it expresses only that part of the genome that it needs for the purpose. Cells are designed economically.

Does economy also prevail in the design of the organism at the macroscopic level? Consider the locomotor system of mammals, in which the basic body plan, the skeleton as well as the different muscle groups, is the same for all species. We have already seen that different uses of the musculature require different parts of the locomotor system to be developed and others reduced. Remember the design of the elegant legs of the pronghorn, where the muscle mass of the proximal hind legs is very pronounced because the main thrust for the pronghorn's fast

Figure 1.4 Blind mole rat *(Spalax ehrenbergi)* from Israel. The radiograph (b) shows a strong forelimb skeleton with a long lever at the elbow for insertion of a greatly enlarged triceps muscle; the hind limbs are relatively small.

run is generated by those legs. Contrast this structure with that of the blind mole rat from Israel shown in Figure 1.4, a fossorial animal that spends its life underground digging. The mole rat performs this work mostly with the head and the front limbs, and accordingly most of the muscle mass is concentrated in the front limbs and the neck; but there is relatively little muscle on the hind limbs, which the mole rat uses only for balance and for steering. Clearly, although all mammals have the same muscle pattern, the muscle mass is well developed where strength is needed and less developed where it is not. Another interesting design difference in these two species is that the antelope has large eyes with which to scan the prairie for predators, whereas the mole rat is blind. Its eye is merely a tiny rudiment that lies beneath sealed lids. This eye can still perceive light-dark cycles, all the animal needs for its underground life, but has lost all the other visual functions, including all the associated structures in the eye and in the brain.

These examples show that both in the cell and in functional organ systems the basic construction plan of mammals is the same, but that it is quantitatively modified to reflect adjustments to specific functional needs—what is often called the optimization of design to serve a specific function. The engineering notion of "optimization" causes difficulties when it is applied to situations that have resulted from the evolution of species through mutations of the genome, because it is rightly said that evolution does not result from engineering. But natural selection would still tend to favor the "best solutions," those that provide the greatest advantage for survival and reproduction.

A highly informative example of optimal or economic design of internal structures is the vascular system. Blood is distributed to the many microvascular units in the tissues, through a sequence of arteries whose diameter is reduced with each branching. What should the diameter of each artery be to ensure efficient blood flow? This basic question was first answered systematically in 1913 by W. R. Hess, later my physiology teacher in medical school, who greatly influenced my thinking in terms of physiology. Hess was awarded the Nobel Prize in 1949 for his remarkable work on the vegetative nervous system; he had worked out the way its two conflicting parts, the ergotropic and the trophotropic systems, interacted to affect body functions. In his earlier analysis of vascular structure, Hess had also argued that two conflicting problems must be solved in order to obtain a vascular system that can transport and distribute blood efficiently and at minimal cost (Figure 1.5): the work needed to pump blood through the vessels depends on the resistance that the vessels offer to the flow, and this resistance is inversely proportional to the fourth power of the vessel radius, according to

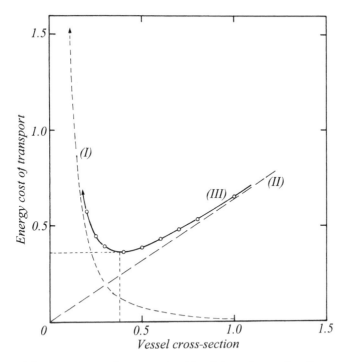

Figure 1.5 The energy cost of transporting blood through vessels of increasing cross-section depends on falling flow resistance (I) and on the increasing mass of blood and of vessel wall that must be maintained (II); the combination of the two factors results in a total cost (III) that is a minimum at optimal vessel size. (From Hess 1913.)

Poiseuille's law; the cost of this work thus becomes smaller the larger the vessel (Figure 1.5, I). Large vessels are costly for at least three reasons: they require a large volume of blood, their walls need a large mass of tissue, and they occupy much space; as a result the cost increases linearly with vessel size (Figure 1.5, II). Hess hypothesized that, in all parts of the arterial tree, there should be a vessel size where the total cost, the sum of (I) and (II), is minimal, and he obtained the graph shown in Figure 1.5 (III), which indeed shows a minimum value. From this he then predicted that the size of the vessels must be reduced at each branch point by a factor of cube root of ½, which is, indeed, what one finds, as we shall see in greater detail in Chapter 6.

Such examples suggest that structural design is finely regulated to the function served and that this regulation follows a cost-to-benefit relationship: the larger an organ the greater its functional yield, but also the higher its cost in terms of construction, maintenance, and load on the body. One of the strongest pressures or constraints on keeping each unit as small as possible is the competition for space, because the composition of the body resembles that of a very crowded city: there is standing room only.

The Principle of Symmorphosis

I have now described three basic principles that determine the design of biological organisms, principles that underlie how cells, tissues, and organs are made: adaptation, integration, and economy. Inasmuch as structural design is concerned, these three principles are, in fact, hardly separable: economic design implies adaptation of structure to functional need and can only be determined if integration of the parts into the whole is considered.

This inseparability of the three principles became apparent when, in 1980, C. Richard Taylor and I analyzed the results of a large study on the respiratory system of mammals in which Taylor's group and my own team had been involved for several years (see Chapter 8). We discussed integration of function from the lung to the mitochondria, as well as economy of design if "animals were built reasonably"—as we surmised—and we considered the role of adaptation of design to functional need. We then realized that we were, in fact, discussing different facets of one and the same principle, which we named "symmorphosis." The root of this new term is μορφωσις, Greek for formation; the Greek prefix συν– signifies something like "balanced formation," similar to the meaning "balanced measures" in "symmetry." We originally defined symmorphosis as a "state of structural design commensurate to functional needs resulting from regulated morphogenesis whereby the formation of structural elements is regulated to satisfy but not exceed the requirements of the functional system" (Taylor and Weibel 1981). It is immediately obvious that the principles of adaptation, integration, and economy are satisfied if structural design is commensurate to functional needs throughout the organism.

It is intuitively plausible that symmorphosis is particularly important when we are considering not individual organs but functional organ systems made of many connected parts, such as the system for energy supply to the body. According to symmorphosis, the size of the parts must be matched to the overall functional demand; in other words, the quantitative design of the parts must be adjusted to the quantitative overall functional needs, with costs and benefits being balanced on a broad scale—design must be optimized.

The principle of symmorphosis is admittedly quite radical. It postulates, in effect, precise tuning of all structural elements to each other and to the overall functional demand. This type and degree of coadjustment of structures is perhaps an unlikely outcome of evolution by natural selection, as has been rightly remarked, if only because functional demand may be variable. Furthermore, if we are studying the system in animals as they appear today, the conditions may now well be different from conditions in the distant past when selective pressure favored particular design properties that were adaptive at the time. But this limitation may apply mostly to traits that interface with the unpredictable environment, whereas traits that characterize the complex internal constitution of the body may not be affected significantly by environmental change. This is therefore not sufficient reason to reject symmorphosis a priori.

A basic notion of symmorphosis is that animals must provide their complex systems with a *functional capacity* that can cope with the highest expected functional demands. So symmorphosis is essentially dealing with capacity adaptation. Since all functional systems are made of many connected parts we must postulate that each part must be made commensurate to the overall load on the system, the basic postulate of symmorphosis. A functional system is like a chain made of many links, which is said to be as strong as the weakest link. To make the chain of a given strength all links must therefore be made equally strong; it is useless—and wasteful—to make one link much stronger. However, risk/cost factors and their uncertainties must also be taken into consideration. Since the strength or stress resistance of each link is not precisely determined but shows some statistical variability, "cheap" links can well be provided with a greater safety factor than "expensive" ones, as in the case of a chain made of different materials of different

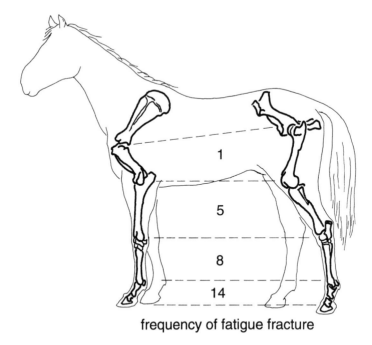

Figure 1.6 The tendency to optimize the design of the leg bones of the horse for running by making them slimmer results in an increased frequency of fatigue fractures toward the periphery. (From Alexander 1998.)

price, gold and iron for example. In this case it may be in the interest of overall economy to make the gold link a little bit less strong than the others, thus decreasing cost and taking a slightly greater risk.

That this process occurs in nature is shown by the structure of the legs of racehorses, well described by R. McNeill Alexander in 1998. As in the case of the pronghorn, the peripheral leg bones of the horse are less strong than the proximal ones (Figure 1.6). This is a cost-saving adaptation because as the legs swing backward and forward, their lower parts must be accelerated through much larger ranges of velocity than their upper parts. Therefore, having the bones of the lower parts of the leg stronger would put more burden on the muscles, and we find the cannon bones built with smaller cross-section and hence with a lower safety factor than the bones higher up the leg—a paradoxical solution because the peripheral bones carry greater weight. As a result, fatigue fractures are more common in the long cannon bones of the horses' feet than in the bones higher up the leg. Alexander argues that this saving on bone tissue in the peripheral bones is reasonable because the stresses in a bone are quite variable, since the strength of the bone is to some extent unpredictable. Variable safety factors can, of course, introduce uncertainties when we attempt to work out the quantitative relationships between structural components and overall functions. Possible safety factors must therefore be accounted for in the analysis.

If we propose that symmorphosis is the principle governing the design of functional systems that leads to coadjustment of their parts to the overall function, we can then set up quantitative (physico-mathematical) models of how structure and function are related and use symmorphosis to make predictions about how the parts must be designed quantitatively to accord with the theory. These predictions can then be used as hypotheses and can be subjected to experimental tests. They can be either accepted or refuted; in the latter case we can see whether there is an alternative interpretation of the findings that leads to a modified hypothesis.

In the following chapters I will use this approach to work out quantitative relations between form and function. I will focus on the functional system of energy supply to the muscles, mainly because energetics is an eminently vital function that is measurable, and because energy supply involves multiple organs such as the lung, the gut, the liver, the heart, the blood, and capillaries for the supply of oxygen and fuels to the cells. The energy-supply system even reaches deep into the cells themselves, because the central organelle of energy production, the mitochondrion, is a constituent of all cells. Furthermore, the production of energy for cell work can be related to other kinds of work—mechanical work in the case of muscle—and it has a measurable limit, so that the overall functional capacity of the system can be defined. We can then attempt to relate the design characteristics of the structures involved in energy supply to this functional limit. A further justification for focusing on muscle and the energy-supply system, if such justification is needed, is that this complex organ system occupies the major part of the body.

The approach I will take is to proceed systematically from the level of the cell up to tissues, organs, and finally organ systems for energy supply. The case studies will be theory driven, since I will work out a hierarchy of consistent quantitative models and make predictions on the basis of the theory of symmorphosis that will then be tested. I will use mainly variations in the design and performance of the system as they result from adaptation to different life styles and environmental conditions (adaptive variation) or from body size (allometric variation).

In brief, this is what is required for the test of the theory of symmorphosis:

1. We first need to assess the functional capacity of the system by measuring the limit of functional performance. In the case of the energy pathway one such limit is the maximal rate of oxygen consumption, $\dot{V}_{O_2}$max, which can be elicited by very strenuous muscle work.

2. We need to set up bioengineering models that define the quantitative effects of design on functional performance, and doing this should allow us to make testable predictions on how each design element should affect functional performance.

3. Just as all physiological information must be quantitative, the design parameters must be measured by morphometry and then introduced into the models to estimate their effect on functional performance.

4. In order to test for symmorphosis, we will make use of the diversity of animals that nature offers. Energetic needs vary because of adaptive variation: horses, dogs, and pronghorns have higher energy needs than cows and goats. Energetic needs also vary because of allometric variation: small animals have relatively higher energetic needs than large ones. One of the crucial tests will therefore be to see whether design parameters show the same type of relation to functional performance when they are studied in cases of allometric and adaptive variation.

Along the way I would like to convey several messages relating to the study of form and function. To physiologists I wish to say that the integration of function depends to a very significant degree on structural design. I also want to show physiologists that morphological information can be as quantitative as physiological data and can thus be a valuable complement in their studies.

The message I wish to send morphologists—my fellow anatomists—is that research on structural design can be rewarding beyond the meticulous descriptions of structural details supported by beautiful micrographs, because advanced integrative physiology crucially depends on adequate information on structural design. But the required structural information must be rigorously quantitative and testable. Qualitative descriptions are no longer enough; morphology must become morphometric.

In general I want to convey some of the fascination I draw from the systematic study of form and function for understanding how animals as diverse as shrews, pronghorns, dogs, goats—and humans—are so different though made on the basis of essentially the same blueprint. Much hidden beauty can be uncovered by the systematic search for the relations between form and function.

2: Cells and Tissues: Oxidative Metabolism in Muscle

All eukaryotic cells are built on a basic plan that establishes the order in which different cell functions are performed under suitable conditions. The main structuring elements are lipoprotein membranes that separate different compartments. A plasma membrane bounds the cell at its surface; it comprises a large array of proteins that are inserted in the lipid bilayer and form different functional entities, such as channels for ion and solute exchange, proteins for anchoring other cell structures, adhesion molecules with which the cell can attach to outside structures or other cells, and receptors for signaling molecules or antibodies. The lipid bilayer makes this membrane practically impermeable to ions or aqueous solutes, establishing a strong electrolyte gradient between the inside of the cell and the interstitial fluid by the activity of specific ion pumps, such that the inside is potassium-rich whereas the outside is abundant in sodium. This gradient establishes an electrical potential across the membrane; in nerve and muscle cells transient changes in this potential are used to transmit signals that trigger cell activities.

The cell itself is subdivided by a special perinuclear membrane system into the cytoplasm and the nucleus, which houses the DNA of the genome and the enzyme machinery for replication and transcription of genetic information from DNA to RNA. The nucleus is very similar in all cell types, whereas the structure of the cytoplasm is highly variable, depending on the cell's specific task. Nevertheless, there is a basic structure of the cytoplasm that is modulated in the various cell types in a typical way. The cytoplasm is subdivided into subcompartments by a complex cytoplasmic membrane system, consisting of membranes thinner than the plasma membrane but similar in basic construction. They form the endoplasmic reticulum (ER) as a set of tubules, vesicles, and flat cisternae, sites of many metabolic functions; the endoplasmic reticulum is related to the membrane complex called the Golgi apparatus, which mediates connections between the ER and the plasma membrane as well as connections with other membrane-bounded organelles such as lysosomes or secretion granules. Many metabolic functions take place in the actual cytosol, the seemingly unstructured space between the different membrane systems. The cytosol contains as its main organelle the ribosomes, through which the genetic information carried by messenger RNA from the nucleus is translated into the polypeptide chain of proteins; ribosomes that make proteins for export or for insertion into membranes become attached to the flat cisternae of the ER, from where the proteins reach their targets via the Golgi complex, which serves as a sorting center.

Additional characteristic constituents of the cytosol are filaments and fibrils that build a cytoskeleton and are responsible for the mechanical stability and motility of the cell. In essence the cytosol of all cells is pervaded by a mesh of fine filaments which extend from membrane to membrane and to which the endoplasmic membranous components, such as vesicles, are attached. Not only do these filaments establish order within the cytoplasm, but they also serve as transfer ropes along which vesicles are moved, for example, from the Golgi complex to various sites. This movement is effected by a small protein that attaches the vesicle to the filament. The major constituent of these filaments is actin, which interacts with myosin to form a complex that effects movements. This occurs in nearly all cells to a certain extent; in cells such as macrophages that move about the tissue, a mass of actin and myosin occupies the cell space at the leading edge of the moving cell; this mass can deform the cell periphery by contraction and relaxation and thus push the leading edge forward. This is a very ancient mechanism, found even in amoeba. The complex of actin and myosin filaments reaches its highest degree of differentiation in the contractile matter of muscle cells, about which I will say more below.

The cytoplasm also contains a special type of organelles I will discuss in great detail: the mitochondria, which are responsible for energy production by oxidative phosphorylation; they are the little furnaces of the cell. Mitochondria possess two concentric membranes and contain their own circular DNA, which codes for part of the mitochondrial enzymes. It is believed that mitochondria are endosymbiotic organelles representing a special bacterium that has been incorporated during evolution of the eukaryotic cell.

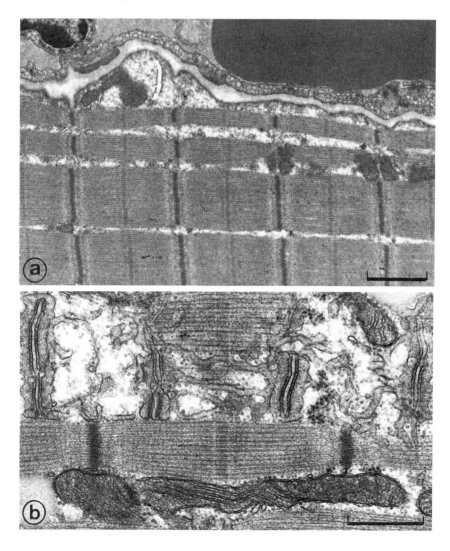

This very general description of cell structure and basic cellular functions will suffice for my purposes. In the course of development and differentiation, each cell expresses in adequate quantity those proteins that are required to serve its assigned functions. If the proteins are incorporated in the membranes of the endoplasmic reticulum, these cells will build a large complement of ER membranes, such that the membrane surface is quantitatively adequate for the cell's required

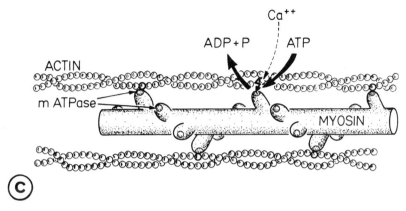

Figure 2.1 The contractile units of skeletal muscle. The electron micrograph of a muscle cell in (a) shows the contractile myofibrils with their cross-bands formed by the alternating and overlapping arrangement of the two filaments actin (thin) and myosin (thick), as shown in detail in the electron micrograph in (b) and schematically in (c). In (a) a capillary is seen lying outside the cell membrane; it supplies the muscle cell and its mitochondria with fuels and oxygen for the production of ATP as the energy currency for actin-myosin interaction (c). The tubular membrane system called the sarcoplasmic reticulum, seen in (b), controls contraction through the release and recovery of calcium ions, as seen in (c), which reveals the arrangement of myosin heads and their relation to actin filaments. Scale markers: (a) 1 μm, (b) 0.5 μm.

function. Examples of such adjustment of cell structure to function are many.

Consider as an example of the adjustment of cell function the skeletal muscle cell, and ask whether the design of such cells reflects their potential for functional performance. Muscle cells have the same basic constitution as other cells with one important exception: they form very long fibers and contain many nuclei so that the many "cells" that form a fiber are not separated by plasma membranes. The whole fiber is, however, enwrapped by a plasma membrane that is here called the sarcolemma.

Muscle cells have one chief function: to produce a forceful movement by shortening. In order to achieve this mechanical work they are provided with myofibrils made of two sets of filaments, actin and myosin, which are organized in sarcomeres (Figure 2.1); these filaments are special derivatives of the cytosolic filament system that, in all cells, is also constituted to a large extent of actin associated with myosin. In skeletal muscle, actin and myosin are of a special type—muscle myosin even occurs in several isoforms that determine the velocity of contraction; they form filaments that lie alongside each other in a regular array within the sarcomere (Figure 2.1a). The shortening is achieved by the sliding of the myosin along the actin filaments. This is performed by a row of little "heads" or "feet" formed on the myosin that are sometimes called the myosin motors: these feet attach to the actin in a cyclical motion to form cross-bridges, and while bending they pull the myosin along the actin filament, a process that is similar to a caterpillar moving along a string by shifting its feet forward one after another.

The movement of myosin feet requires a signal and energy. Contraction is triggered by an electrical signal elicited by the discharge of a motoneuron nerve fiber on the plasmalemma that spreads along the fiber as an action potential. This signal is carried into the cell by tubules of the plasmalemma, the t-tubules, which then transmit the signal to tubules of the endoplasmic reticulum, here called sarcoplasmic reticulum, or SR (Figure 2.1b). The membranes of SR are equipped with a calcium channel that releases Ca^{2+} ions into the myofibrils, thus triggering the movement of myosin heads. It is noteworthy that the SR is richly deployed between the myofibrils so that it can pump the Ca^{2+} ions back into the membrane space, a process that requires energy in the form of adenosine triphosphate (ATP).

The shortening of the muscle fiber results from the summed shortening of the many sarcomeres of the myofibril arranged in series to form a long chain from one end of the fiber to the other. The force generated is proportional to the number of parallel myofibrils that constitute the contractile machinery of the muscle cell. All in all, the work performed is proportional to the total number of myosin heads active.

This process must be fueled by energy, mostly for operating the myosin motors but also for pumping Ca^{2+} back into the SR; the energy for the latter is about proportional to the energy consumed by the myosin motors so I will not always treat it separately. One of the important properties of muscle cells is, therefore, that they can generate the energy at the rate required by the myosin motors and the SR. This energy comes from the high-energy phosphate ATP: when a myosin head detaches from the actin in order to "move ahead," the required energy is obtained by splitting off one phosphate group from ATP (Figure 2.1c), the myosin head acting as an ATPase, the enzyme splitting ATP. This results in adenosine diphosphate, ADP, which must be recharged to ATP in the muscle cell itself because ATP cannot be imported into the cell from outside. We shall see that the cell possesses two mechanisms to do this: one anaerobic, which does not require oxygen, and another aerobic, which can only function if O_2 is supplied at an adequate rate.

In the following discussion of the relation of form and function in the muscle cells I will concentrate on the mechanisms for generating this energy and the equipment required, for several reasons: (1) The energetic needs of muscle cells are related to mechanical work, which is easily measured either in the muscle or in the whole organism. (2) For prolonged work, muscle must generate the high-energy phosphate ATP by oxidative phosphorylation, and therefore muscle work can be measured by oxygen consumption. (3) The structures involved in oxidative phosphorylation, the mitochondria, are well known and measurable by morphometric techniques. (4) Muscle is the largest organ mass in the body, constituting about 40% of body mass, and when muscles are put to work, over 90% of the energy consumption by the body occurs in muscle. Indeed, it seems that the energy-supply system of the body is gauged to muscle work, for only strenuous muscle work can stress the energy-supply system of the body to its limit of performance. Starting with the energy metabolism of muscle cells therefore lays the groundwork for extending our study, in subsequent chapters, to the pathways for oxygen and fuel supply to the body.

Energy Supply and Mitochondria

How is a steady and sufficient supply of ATP to actomyosin ensured? The cell stores a small pool of high-energy phosphate in the form of creatine phosphate, CrP, which immediately restores ATP near the myosin-ATPase by donating its phosphate group to ADP, a process which is catalyzed by the enzyme creatine phosphokinase (Figure 2.2). But this pool is very limited and suffices only for a few seconds of muscle work, a jump or a short sprint. Work of longer duration depends on replenishing the high-energy phosphate pool through metabolism of fuels, for which two pathways are available, one anaerobic and

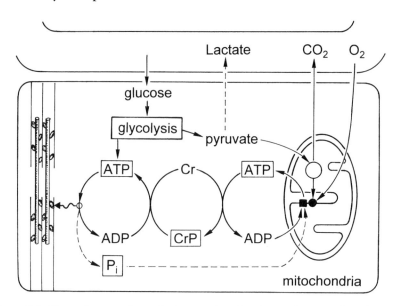

Figure 2.2 Flow of energy through the muscle cell. Glucose from capillary blood is broken down in glycolysis. The product pyruvate either is released as lactate (anaerobic condition) or it is shuttled into the mitochondria for oxidative breakdown to carbon dioxide and water. The ATP resulting from these processes is carried to the myofilaments via a creatine phosphate shuttle system, where the enzyme creatine phosphokinase catalyzes the binding of phosphate, P, to creatine, Cr, and its release.

the other aerobic, that is, dependent on the oxidative combustion of fuel stuffs. This is explained in some detail in Box 2.1. In short, these pathways are as follows:

The anaerobic breakdown of glucose to pyruvate by the enzyme system of glycolysis generates some ATP; this is a fast process but it is quite inefficient because it uses 1 mol glucose to produce 2 mol ATP, and the end product is lactic acid, which is discharged into the blood with the result that blood pH falls, a condition that is not well tolerated by the body. Anaerobic glycolysis is therefore mainly used for fast burst contractions, or when aerobic metabolism is not sufficient to provide the energy needed.

The more efficient mechanism is oxidative phosphorylation, which allows 36 mol ATP to be formed from 1 mol glucose. This process occurs in the mitochondria. It uses as fuel either glucose or fatty acids.

Box 2.1 Oxidative and anaerobic phosphorylation in muscle cells

Anaerobic ATP production occurs in the cytoplasm. But the metabolic pathway that leads to the production of ATP through oxidative phosphorylation of ADP is largely housed in the mitochondria (Figure B2.1). It begins with the breakdown of fuels or substrates in the cytoplasm by different mechanisms for the different substrates. This leads to the common pathway of oxidative metabolism, which begins with the Krebs tricarboxylic acid cycle and ends at the respiratory chain complex (Figure 2.6a).

The mitochondrion is made of two concentric membranes (Figure 2.5). The outer membrane forms the boundary toward the cytoplasm. The inner membrane is apposed to the outer membrane but forms many infoldings, the cristae mitochondriales, which carry the respiratory chain complex and the F_1-ATPase that is responsible for phosphorylation. The inner membrane forms the boundary of the mitochondrial matrix, which houses the Krebs cycle enzymes. The intermembrane space between the outer and inner membranes is generally very narrow and extends into the cristae.

The *final common pathway* of oxidative metabolism begins at the Krebs cycle with the introduction of a 2-carbon acetic acid from acetyl-CoA, which condenses with oxaloacetic acid to make citric acid. Through several enzymatic steps these two carbons are clipped off and released as two CO_2 molecules that diffuse into the blood. As this happens, four protons H^+ (reducing equivalents) are donated to the electron acceptor NAD to make NADH and four additional H^+ are obtained from two molecules of H_2O that are used in the process of making the two CO_2 molecules. The reducing equivalents (NADH + H^+) are supplied to the respiratory chain in the cristae of the inner membrane, where they are transferred through an electron transport chain of various enzymes, finally being oxidized to H_2O at the terminal oxidase and thus consuming half the O_2 molecule that is supplied from the blood. As the reducing equivalent travels along the respiratory chain, it gradually loses energy, which is used to establish a proton (pH) gradient across the inner membrane (chemiosmotic theory), which is utilized for making three ATP molecules on the F_1-ATPase bound to the inner membrane (see Figure 2.6) for each proton transferred.

The *entrance of acetate* into the Krebs cycle occurs from acetyl-CoA

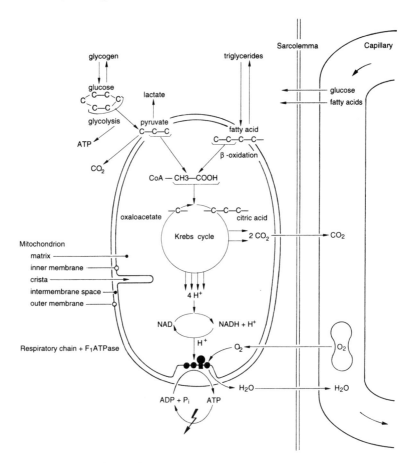

Figure B2.1 Pathways for substrate metabolism in mitochondria.

which is obtained by catabolism of substrates along different pathways. In the *carbohydrate* pathway, glucose (which may be derived from glycogen) goes through glycolysis in the cytoplasm and is broken into two pyruvate molecules while two ATP are formed as well as NADH, which must be reoxidized to NAD in the mitochondria. Glycolysis is a fast anaerobic reaction. Its end product, pyruvate, has two fates: (1) under anaerobic conditions it becomes reduced to lactate by accepting an H^+ from NADH, which thus becomes oxidized; or (2) under aerobic conditions it enters a mitochondrion where it donates a 2-carbon acetyl unit to Coenzyme A while splitting off the third group as CO_2. Acetyl-CoA then supplies the Krebs cycle with an acetate group. In the *fatty acid* pathway, fatty acids are cut from triglycerides in the cytoplasmic lipid droplets and transferred into the mitochondrion by the carnitine shuttle in the mitochondrial membranes. In the mitochondrial matrix the fatty acids enter the process of β-oxidation, where a set of enzymes clips off 2-carbon residues, which combine with Coenzyme A to enter the Krebs cycle as acetyl-CoA.

Oxidative phosphorylation depends on an adequate supply of O_2 from the red blood cells in capillaries and on the removal of the waste product CO_2 by the blood. It also depends on the supply of substrates, partly from intracellular stores but eventually also from the blood. The stoichiometry of the different fuel pathways is as follows: each mole of glucose catabolized consumes 6 mol of O_2, and generates 36 mol of ATP, 6 mol of CO_2, and 42 mol of water. In turn, the oxidation of one mol of fatty acid, made of a 16-carbon chain, generates 130 mol ATP consuming 23 mol O_2 and producing 16 mol CO_2.

. . . .

In the cytoplasm, glucose must first go through anaerobic glycolysis, and pyruvate donates 2-carbon acetate to the mitochondria. Fatty acids are also broken stepwise into acetate residues, but this process occurs in the mitochondrial matrix. We shall see in greater detail that the acetate is oxidized and completely broken down to CO_2 and water, much cleaner wastes than lactic acid because they can be eliminated from the blood very easily through the lungs and the kidneys. The energy gained by the process of oxidation of fuels is transferred to ATP.

As a result of these differences in the two main pathways for generating ATP, endurance exercise can be supported only by oxidative phosphorylation because acidosis, resulting from the accumulation of lactic acid from glycolysis, limits work duration. But this requires that O_2 be supplied at an adequate rate from the capillaries (Figure 2.2) because the cells cannot store O_2 in appreciable quantity. Figure 2.3, taken from the classic work of Rodolpho Margaria (1963), shows that O_2 consumption increases linearly in individuals running on a treadmill when running speed increases, demonstrating that the energy used to fuel the contractile machinery of muscle cells is proportional to the work output of the muscles. The rate of O_2 consumption is therefore a measure

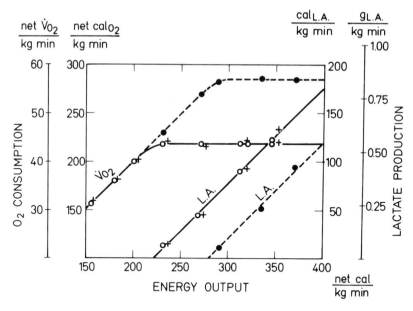

Figure 2.3 Oxygen consumption ($\dot{V}_{O_2}$) in running humans increases with exercise or work intensity (energy output) up to a maximum, where it reaches a plateau; after that, higher work rates are fueled by anaerobic glycolysis, which results in an increase in lactate concentration (L.A.) in the blood. Maximal oxygen consumption varies with individuals with athletes having higher values than nonathletes. (From Margaria et al. 1963.)

of total energy supply to the muscle cells. But O_2 consumption reaches a plateau at a certain point, beyond which a further increase in running speed is not fueled by increased oxygen consumption. The additional energy required is instead covered from glycolysis: large amounts of glucose are metabolized anaerobically and accordingly we find that lactic acid begins to accumulate in the blood at an increasing rate. From this we can draw two important conclusions: (1) the muscle cells have a well-defined capacity for oxidative metabolism which can be measured as the maximal O_2 consumption, $\dot{V}_{O_2}$max at the plateau in Figure 2.3; (2) the rate of oxidative metabolism is limited not by the work capacity of the muscle myofibrils, but rather by the machinery supporting oxidative phosphorylation—that is, by processes essentially occurring in the mitochondria—and by O_2 supply from the capillaries.

This account provides a good starting point for our first test of symmorphosis. Consider the hypothesis that functional capacity is related to structural design at the level of cells and tissue. First, since oxidative phosphorylation occurs exclusively in the mitochondria, we predict that the amount of mitochondria is proportional to the capacity for oxidative phosphorylation estimated by $\dot{V}_{O_2}$max. Second, since oxygen must be steadily supplied from the erythrocytes in the capillaries because the muscle cells cannot store oxygen, we predict that the capillaries are quantitatively matched to the oxygen needs of the muscle cell mitochondria, that is, proportional to $\dot{V}_{O_2}$max. We will test these predictions by comparing individuals with different $\dot{V}_{O_2}$max. Figure 2.3 shows that some subjects reach $\dot{V}_{O_2}$max at higher levels than others; so the question arises whether they have proportionately more mitochondria and more capillaries.

Is Mitochondrial Structure Matched to the Demand for Oxidative Energy?

Before we can address this question we must ask how the functional process of oxidative energy production is related to mitochondrial structure. What are the parameters that relate form to function in this subcellular organelle? This requires that we first have a look at mitochondrial structure and at the biochemical processes that break down fuels, mainly glucose and fatty acids, in order to liberate the energy that is transferred to ATP.

In muscle, as in all cells, mitochondria are small cytoplasmic organelles that are bounded by a membrane (Figure 2.4). In muscle cells they are found either interspersed between the myofibrils or in small packets beneath the sarcolemma. Their diameter is on the order of 1 μm. Their length, however, is indeterminate, because their sausage shape is typically quite complicated, forming networks extending through large parts of the muscle cell.

The internal organization of mitochondria is of great importance for understanding their function. Figure 2.5 shows an electron micrograph of a muscle mitochondrion, where we see that this organelle is typically made of two membranes and two spaces. The *outer mitochondrial membrane* simply looks like an envelope around the organelle that separates it from the surrounding cytoplasm. It is more or less smooth

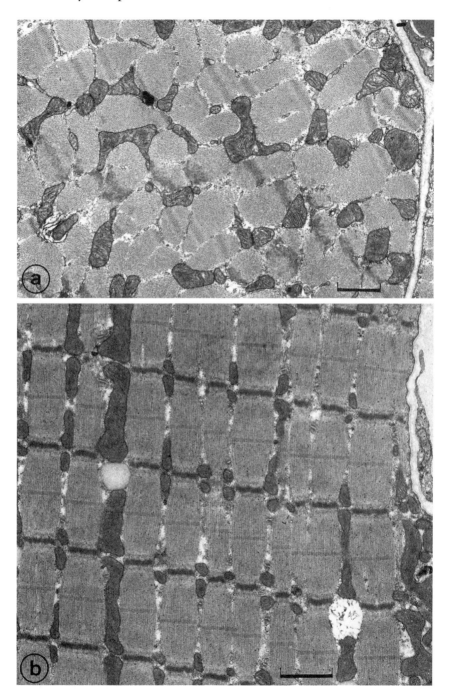

and defines the mitochondrial space. The *inner mitochondrial membrane*, in contrast, is partly apposed to the outer membrane but forms many folds, the mitochondrial *cristae*, with the result that the inner membrane area is several times larger than the surface area of the outer membrane. The *intermembrane space* extends into the cristae and is believed to be very narrow in vivo (it is somewhat widened in electron micrographs prepared by standard methods). The innermost compartment is the mitochondrial *matrix*, which in electron micrographs appears as a finely granular structure. It is completely bounded by the inner mitochondrial membrane.

Mitochondrial structure is closely connected to the oxidative breakdown of fuels, a multistage process (Box 2.1). In brief, the carbohydrate glucose as well as the hydrocarbon chains of fatty acids are first broken into 2-carbon acetic acid, which is bound to Coenzyme A (CoA) in order to be introduced into the Krebs cycle. This cycle is a circular chain of enzymes that gradually break the acetic acid into two molecules of CO_2, and this process liberates four protons (H^+). These are transferred, one by one, to the respiratory chain, a complex of enzymes that transfer the hydrogen ions stepwise to the last enzyme, the cytochrome oxidase a_3, where they are oxidized to water as the end waste product, consuming molecular oxygen that diffuses in from the blood. In the stepwise process of proton transfer the enzyme complex can extract the energy liberated and transfer it to a linked enzyme, the F_1-ATPase, which generates ATP. It is known from many biochemical studies that this process is highly efficient and also tightly controlled such that for each mole of O_2 consumed, 6 mol ATP are generated. Looking at the fuels, we know that each mole of glucose catabolized consumes 6 mol of O_2, and generates 36 mol of ATP, whereas the oxidation of 1 mol fatty acid, made of a 16-carbon chain, consumes 23 mol O_2 and generates 130 mol ATP. These processes are very efficient; about 40% of the energy contained in the fuel is transferred to ATP.

Where are these different steps localized in the mitochondria? It is

Figure 2.4 Electron micrographs of transverse (a) and longitudinal (b) sections of muscle cells show that mitochondria are interspersed between myofibrils and have a complex shape, appearing branched mainly on transverse sections. Scale markers: 1 μm.

38 · · · Symmorphosis

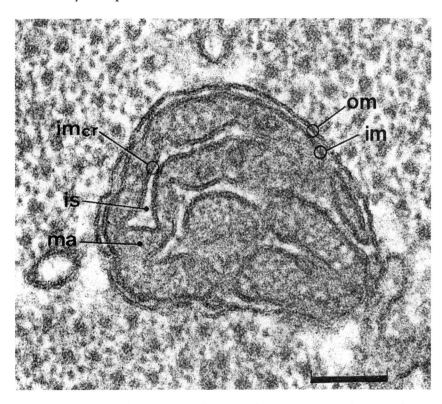

Figure 2.5 Mitochondria are made of two membrane systems, as shown on this electron micrograph of a muscle cell: the outer membrane (om) and the inner membrane (im), which separates matrix (ma) and intermembrane space (is) and forms cristae, thus increasing the interface area between matrix (where the Krebs cycle enzymes are housed) and the inner membrane that carries the complex of respiratory chain enzymes coupled with the F_1-ATPase as shown in the models of Figure 2.6. Scale marker: 0.1 μm.

now well established that the mitochondrial matrix houses the entire enzyme system for the Krebs cycle as well as that for β-oxidation of fatty acids (Box 2.1). The enzyme complex of the respiratory chain, however, is built into the inner membrane together with the F_1-ATPase (Figure 2.6a). The chain comprises four enzyme complexes which transfer electrons stepwise to the terminal complex IV, the cytochrome oxidase a_3, where finally O_2 consumption allows the protons H^+ to be oxidized to H_2O, the final product of mitochondrial oxidation. These enzymes are built into the lipid layer of the membrane as shown in

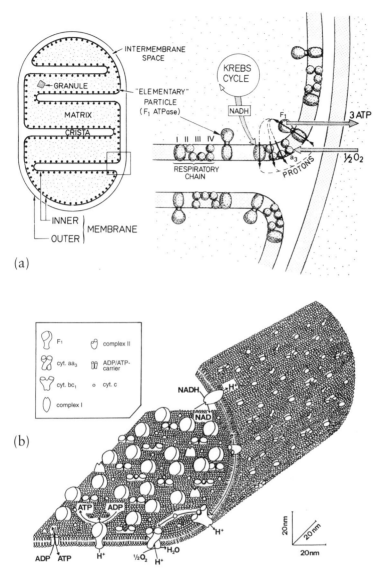

Figure 2.6 (a) Model of mitochondrial structure showing molecular organization of the respiratory chain and F_1-ATPase. Electron transfer along the respiratory chain causes proton flux from matrix to intermembrane space at the three complexes I to III and is terminated at the terminal complex IV with cytochrome oxidase a_3 where oxidation occurs; the proton gradient thus established drives phosphorylation of ADP to ATP on the F_1-ATPase. (From Weibel 1984.) (b) Molecular model of inner mitochondrial membrane showing the insertion of the different protein complexes into the phospholipid bilayer. Note that the complexes are not arranged as a physical chain, as in the scheme of Figure 2.6a. (Courtesy K. Schwerzmann.)

Figure 2.6b. It is interesting that the concentration of these complexes in the membrane is reasonably balanced, except for the cytochrome oxidase a_3, which occurs in more than double concentration. It has recently been shown that this apparent excess capacity in this critical enzyme, the terminal oxidase, is most important for the regulation of a high overall mitochondrial O_2 affinity, that is, for maintaining the optimal functioning of the entire respiratory chain. The molecular design of the respiratory chain is therefore well balanced.

What are the relevant structural parameters that characterize the oxidative capacity of mitochondria? If we choose the maximal rate of oxidation, $\dot{V}_{O_2}$max, as the functional reference parameter, then the most relevant structure is the inner mitochondrial membrane. Many studies combining biochemistry, cell physiology, and electron microscopy have provided ample evidence that the respiratory chain complexes are densely packed into this membrane; in fact, it is little more than these enzyme proteins with a small complement of phospholipids that form the structural backbone of this membrane and serve as a tight barrier to the passage of charged molecules or ions (Figure 2.6). In essence, this barrier function of the inner membrane is the basis for the chemiosmotic theory of oxidative phosphorylation formulated by Peter Mitchell to explain how the energy harvested from fuels is transferred to ATP. The structural parameter tightly related to the number of respiratory chain units and F_1ATPases is, therefore, the *surface area of the inner mitochondrial membrane,* S(imi).

The second important compartment is the matrix space, for which we may postulate two important structural characteristics: the volume of the matrix space must be sufficiently large to house the set of enzymes required for the Krebs cycle and for β-oxidation of fatty acids, and it must be patterned in such a way as to keep the distance from the Krebs cycle enzymes to the inner membrane as short as possible, to ensure an intimate spatial relation between the proton generator and the site of combustion. This latter requirement is solved by enfolding the inner membrane in the form of cristae so that the matrix space is associated with a large surface for interaction with the respiratory chain. Indeed, the "layer" of the matrix associated with the inner membrane is only about 20–30 nm deep, a distance sufficiently short for enzyme interaction.

Combining these two requirements, we see immediately that the most basic structural parameter of mitochondrial design is the *surface density of the inner mitochondrial membrane within the mitochondrial volume,* which we symbolize as $S_V(imi,mi)$; it directly estimates the relationship between the inner membrane surface and matrix volume. This parameter is not easy to measure on standard electron microscopic sections, but with the aid of special morphometric methods it was found that $S_V(imi,mi)$ is about 35 $\mu m^2 \cdot \mu m^{-3}$, which is equivalent to 35 m² of membrane per 1 ml of mitochondria. This is indeed a very tight packing of these membranes, equivalent to packing a thousand-page book into a thimble! But since the inner mitochondrial membrane is only about 6 nm thick, this packing still leaves space for a layer of matrix of about 20–30 nm over the inner membrane. The value of 35 $\mu m^2 \cdot \mu m^{-3}$ is characteristic for all muscle types, and it appears to be invariant in all mammalian species considered, from the mouse to the cow.

The constant relationship between inner membrane and matrix space in mammalian muscle mitochondria tells us that the structural compartments which house the respiratory chain enzymes and the enzymes of fuel catabolism in the Krebs cycle are matched to each other by design.

Testing for a Quantitative Match of Form and Function in Muscle Mitochondria

The hypothesis of symmorphosis predicts that the structure carrying the enzymes for oxidative phosphorylation must be proportional to the functional capacity, $\dot{V}_{O_2}$max. This structure is the total surface area of the inner mitochondrial membrane, $S(imi)$, which is obtained as the product of $S_V(imi,mi)$ and the total mitochondrial volume, $V(mi)$, in muscle cells.

This prediction assumes that all units are recruited and are working near their sustainable maximal rate when muscles reach or exceed $\dot{V}_{O_2}$max. Under these conditions the intracellular pathway from fuel to oxidation (see Box 2.1 and Figure 2.2) becomes limited at the terminal step, and some of the pyruvate generated by glycolysis can now no longer be shuttled into the oxidative pathway but must be converted to lactic acid, as observed when energy output exceeds the oxidative

Box 2.2 Estimation of morphometric parameters of muscle cells

Morphometric parameters of muscle cells and tissue include the volume of mitochondria, the surface area of the inner mitochondrial membrane, the volume of the capillary blood and the proportion of blood taken up by red cells, and the surface areas of the capillaries and of the sarcolemma, the muscle cell's boundary membrane (Figure B2.2). There are two main problems with the estimation of these parameters: (1) the parameters all relate to a three-dimensional structure, but they can be measured only on flat two-dimensional sections; (2) they characterize a large organ mass (weighing a few hundred kilos in a horse), but they relate to microscopic features that can often only be studied with sufficient precision by electron microscopy. Both problems are solved by the methods of *stereology*, a set of mathematical methods that are based on the theory of sampling (Weibel 1979; Cruz-Orive and Weibel 1990).

The first principle of stereology states that three-dimensional objects can be sampled by random flat sections such that the three-dimensional parameters are represented on the section profiles of the object: the volume V is represented by the area A of the object, the surface S by the contour length B, and so on.

The second principle states that measurements are best obtained if they are related to a reference parameter of the spatial structure that is also represented on the section. We can choose the volume of muscle cells as the reference parameter and express the volume and surface of mitochondria as the volume density V_V and the surface density S_V of mitochondria in muscle cells; the precise notation is $V_V(mi,f)$ and $S_V(imi,f)$, where mi stands for mitochondria, imi for inner mitochondrial membrane, and f for muscle fibers (or cells).

The third principle says that V_V and S_V can be estimated very efficiently by making use of further reductions using test grids overlaid on the section (Figure B2.2). Thus the relative volume V_V can be estimated using a set of test points and counting, for example, the points P(mi) falling on mitochondria and all the points P(f) falling on the muscle fibers (the latter include the points falling on mitochondria). We then obtain $V_V(mi,f) = P(mi)/P(f)$ as an unbiased estimate of mitochondrial volume density in muscle cells. Similarly we can use test lines of known length L to estimate the surface by counting the number of

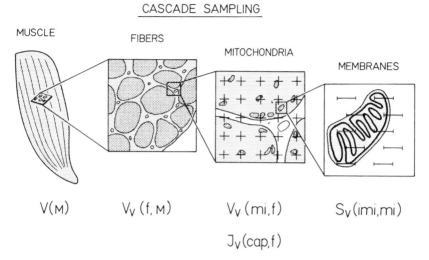

Figure B2.2 Model for morphometric analysis of mitochondria in muscle cells by cascade sampling.

intersections I with the contour of the object of interest on the section. The basic formula is $S_V = 2 \cdot I/L = 2I_L$. If we apply this to the inner mitochondrial membrane expressed as surface density within the mitochondria, $S_V(imi,mi)$, we need to count the intersections $I(imi)$ between the inner mitochondrial membrane trace and the test lines, and to estimate the length $L(mi)$ of the test line within the mitochondrial profile to have $S_V(imi,mi) = 2 \cdot I(imi)/L(mi)$. If the test line is made of short line segments, as shown in the Figure B2.2 (4th panel), $L(mi)$ can be estimated by counting the number of endpoints falling on the mitochondrial profile and multiplying by half the line length.

A fourth principle says that the total structure must be sampled systematically so as to account for the different sampling steps in order to calculate the parameters to represent the entire structure. Employing the principle of cascade sampling shown in the figure, we can perform the detailed stereological measurements at the optimal resolution and still obtain data that are relevant for the entire structure. Thus the total surface of the inner mitochondrial membranes is first obtained on high-power electron micrographs (panel 4) as surface density in mitochondria $S_V(imi,mi)$. At somewhat lower magnification (panel 3), we can estimate the volume density of mitochondria in muscle cells, $V_V(mi,f)$, thus availing ourselves of a more representative sample of the muscle; on low-power light micrographs (panel 2), we estimate the volume density of muscle cells in the total muscle, $V_V(f,M)$. Finally (panel 1), we measure the volume $V(M)$ of the muscle mass of interest. With this sampling scheme we can calculate the total surface of inner mitochondrial membranes in a given muscle or muscle group as

$$S(imi) = V(M) \cdot V_V(f,M) \cdot V_V(mi,f) \cdot S_V(imi,mi).$$

Parameters characterizing the capillaries are obtained by similar procedures based on the same principles (Mathieu et al. 1983). Here we must estimate first the length density of capillaries in the muscle volume, $J_V(c,M)$, by counting capillary profiles per unit area A of section, $Q_A(c,M)$. (We use J as the symbol for "length of curve," and Q for "transsection of curve.") If the sections are "isotropic uniform random," that is, if they are cut in a way that gives all orientations and all positions an equal chance, then $J_V(c,M) = 2 \cdot Q_A(c,M)$; but if we choose to work with cross-sections of muscle (cutting angle 0), the coefficient 2 must be replaced by a factor $c(K,0)$ that accounts for the degree of preferred orientation (anisotropy K) of the capillaries and lies between 1, for a completely parallel course, and 2, for a totally random orientation. This factor has been found to depend on the degree of muscle contraction as the capillaries are more stretched and thus are more parallel to the muscle fiber in extended muscle. An average value found experimentally in our material was about 1.24 for skeletal muscle.

. . . .

threshold (Figure 2.3). Considering the tight link between the matrix enzymes and the respiratory chain, resulting in an invariant surface density of the inner mitochondrial membrane, we can further predict that $\dot{V}_{O_2}$max should also be proportional to the volume of the mitochondria, $V(mi)$, at least in mammalian muscle. To consider $V(mi)$ rather than $S(imi)$ as a parameter relating to $\dot{V}_{O_2}$max has the great advantage that it is much easier to measure than $S(imi)$ (Box 2.2) and can be estimated on a much larger sample of muscle tissue so that the measurement becomes more representative and more accurate. However, this simple relation applies only to mammalian muscle. In other taxa the inner constitution of mitochondria may deviate slightly, though only in quantitative terms. Thus the inner membrane surface density is found to be somewhat lower in reptiles and higher in the flight muscles of hummingbirds. In these instances it is better to relate $\dot{V}_{O_2}$max to the surface area of the inner mitochondrial membrane rather than to mitochondrial volume.

I must, however, introduce a cautionary remark. Since $\dot{V}_{O_2}$max is measured on the whole body, the volume of mitochondria found in all muscles of the body must be used in establishing the correlation. More precisely, we must include in the measurement all muscles that are active in exercise conditions under which $\dot{V}_{O_2}$max is measured. In quadrupedal animals that are running or rather galloping at $\dot{V}_{O_2}$max, the entire locomotory musculature is active, but in bipedal humans who are running or bicycling this may not be the case.

Is $\dot{V}_{O_2}$max Related to V(mi) in Exercising Muscle Cells?

Let us now ask whether muscle cells contain just the necessary amount of mitochondria to achieve their characteristic $\dot{V}_{O_2}$max, as we would predict from the general hypothesis of symmorphosis. This question is not easy to approach directly because we do not have a good estimate of the absolute oxidative capacity of mitochondria, but a first attempt would be to see if muscles that are capable of a higher maximal rate of oxidative phosphorylation also have a proportionately larger mitochondrial volume.

This type of question was first asked around 1970, mainly with respect to working out the characteristics of human athletic performance. Basically the question was: what makes the difference between a high-performance athlete and ordinary people? One of those who asked this sort of question was Hans Hoppeler, at the time a medical student as interested in electron microscopy as in sports. He teamed up with Hans Howald, then the head of the Swiss Federal Institute for Sports Research, who had easy access to some of the top athletes in Switzerland. As a student working in my laboratory Hoppeler also had access to electron microscopy and morphometry. The first question Hoppeler and Howald asked was whether the Swiss Team of Orienteers, highly trained endurance athletes who had just won a silver medal at the world championship, had more mitochondria in their muscle than medical students who served as controls, and whether this was in proportion to their respective $\dot{V}_{O_2}$max.

The arguments for this study ran as follows: if $\dot{V}_{O_2}$max is significantly increased in athletes, then the mitochondria can react in one of the following ways. (1) They can maintain the *number* of enzymes in the respiratory chain but increase the maximal rate at which they can operate; in this case $\dot{V}_{O_2}$max is independent of V(mi). (2) If the *maximal rate* of oxidation by a single unit of the respiratory chain is fixed, then the total maximal oxidation rate can be increased only by increasing the number of enzyme units, again in two ways: (2a) by increasing the surface density of inner membranes within the mitochondria, or (2b) by increasing the total mitochondrial volume in muscle cells if the surface density is invariant.

The procedure was to measure $\dot{V}_{O_2}$max while the subjects ran on a treadmill, as shown in Figure 2.3, and to then obtain by needle biopsy a sample of muscle cells of the lateral quadriceps muscle, one of the main leg extensors. The needle biopsy was processed for electron microscopy, and the volume of mitochondria per unit cell volume, called the mitochondrial volume density $V_V(mi,f)$, was estimated as well as the surface density of inner mitochondrial membranes per unit cell volume, Sv(imi,f), using the methods outlined in Box 2.2. The results of this study are shown in Figure 2.7, which reveals that the volume density of mitochondria in muscle cells of the quadriceps was larger in the athletes than in the less athletic medical students, and this in proportion to their larger $\dot{V}_{O_2}$max. It was further found that the surface area of the inner membrane was also increased in the athletes, and this in strict proportion to the mitochondrial volume so that the surface density within the mitochondria was invariant.

We therefore conclude that the body operates according to option (2b): when its muscle cells are required to achieve a higher $\dot{V}_{O_2}$max they achieve this by making more mitochondria of the same kind rather than by modifying enzyme kinetics or the inner structural makeup of mitochondria. It appears that the relation between mitochondrial form and function is invariant, so that the only option for ensuring higher $\dot{V}_{O_2}$max is to increase the mitochondrial volume. The predictions derived from symmorphosis are fulfilled in this case.

In the meantime several other studies have confirmed the observation that endurance exercise training results in a higher capacity for aerobic metabolism measured by a higher $\dot{V}_{O_2}$max, and that this is related to an

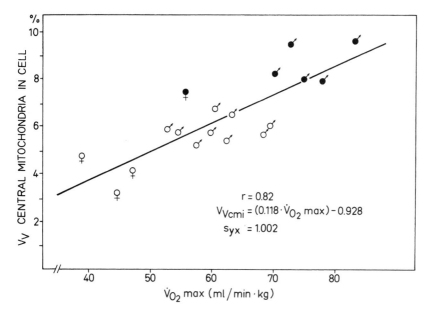

Figure 2.7 The volume density of mitochondria in muscle cells is proportional to $\dot{V}_{O_2}$max which is higher in athletes (black dots) than in sedentary students (open circles). (From Hoppeler et al. 1973.)

increase in the mitochondrial complement of muscle cells. Unfortunately, in many of these studies the "number" of mitochondria is estimated as the form parameter rather than their volume, and this parameter has very little functional meaning. As mentioned earlier the mitochondria have a complex shape, forming vast networks through large parts of the muscle cell. The number of mitochondrial profiles seen and counted on electron micrographs says something about the complexity of mitochondrial shape but is a poor and inaccurate estimator of the quantity of mitochondria present. When attempting to determine quantitative relations between form and function, it is very important to base this work on well-defined parameters, in this case the volume and membrane area of the mitochondria.

Natural Variation in Energy Demand and Mitochondria

The study of athletes just discussed had a number of limitations. First of all, the training of human athletes for higher performance is a very special case that lacks generality. It shows that, in one species and in one specific muscle, mitochondrial volume is increased if energy demand increases. The question must therefore be raised whether the relationship between mitochondrial volume and $\dot{V}_{O_2}$max can also be shown on a broader base—looking, for example, at the variation in energy demand between different species.

Nature has evolved some species that are highly athletic and others that are not. Dogs and horses, for example, achieve a $\dot{V}_{O_2}$max that is 2.5 times higher than that in goats and cows, animal pairs (dog/goat and horse/cow) that can be compared because they are of about the same body size. Man has surely reinforced natural selection by breeding racehorses and greyhound dogs, but their athletic behavior is a natural trait. And horses and dogs do not even represent the extreme. The pronghorn, for example, shown in Figure 1.1, achieves a $\dot{V}_{O_2}$max that is 2 times higher than that of the dog (Table 2.1).

The question arises whether in these athletic species the muscle mitochondria have adapted to higher levels of energy demand by changing their internal characteristics (maximal enzyme rate or density of inner membrane) or by adjusting their volume to be proportional to $\dot{V}_{O_2}$max, as we would predict from symmorphosis.

The comparative study of animal species proceeds along the same lines as the study of humans. The basic reference parameter is maximal oxygen consumption, $\dot{V}_{O_2}$max, measured while the animals run on a treadmill. Muscle is then sampled for microscopic analysis, but in this case we can obtain a large number of samples collected systematically from the whole musculature. Also, we can estimate the total muscle mass, which then allows us to estimate the total mitochondrial volume in the entire locomotor muscle mass and relate it to total maximal oxygen consumption. This is a sound comparison because it is well known that in heavy exercise over 90% of the oxygen consumed is used for oxidative phosphorylation in the skeletal muscles and that furthermore galloping quadrupeds use nearly the total skeletal muscle mass for locomotion. We can therefore expect that nearly all muscle mitochondria will be active when the animal runs at $\dot{V}_{O_2}$max.

Let us now see how the mitochondrial compartment is designed when we compare goats, dogs, and pronghorns, animals that weigh about 30 kg but that differ greatly in their aerobic work capacity in that their $\dot{V}_{O_2}$max per unit body mass M_b is proportional to the ratio

Table 2.1 Morphometry of muscle mitochondria and capillaries of goat, dog, and pronghorn

Parameter	Unit	Goat	Dog	Pronghorn
M_b	kg	27.7	28.2	28.4
M_m/M_b	$g \cdot kg^{-1}$	260.0	370.0	438.0
$\dot{V}_{O_2}max/M_b$	$mlO_2 \cdot min^{-1} \cdot kg^{-1}$	57.0	137.0	272.0
Mitochondria				
$V_V(mi,f)$	%	4.1	8.6	10.6
$V(mi)/M_b$	$ml \cdot kg^{-1}$	10.0	29.7	46.2
$\dot{V}_{O_2}max/V(mi)$	$mlO_2 \cdot min^{-1} \cdot ml^{-1}$	5.7	4.6	5.9
Capillaries				
$J(c)/M_b$	$km \cdot kg^{-1}$	244.0	453.0	653.0
$V(c)/M_b$	$ml \cdot kg^{-1}$	3.9	7.2	10.4
Hct	%	29.9	50.3	45.6
$V(ec)/M_b$	$ml \cdot kg^{-1}$	1.16	3.63	4.74
$\dot{V}_{O_2}max/V(ec)$	$mlO_2 \cdot min^{-1} \cdot ml^{-1}$	49.1	37.7	57.4

Source: The data for dog and goat are from Hoppeler et al. (1987), Conley et al. (1987), and Vock et al. (1996). The data for the pronghorn are from Lindstedt et al. (1991).
Note: M_b, M_m are body and total muscle mass; $V_V(mi,f)$ is volume density of mitochondria in muscle fibers; $V(mi)$, $V(c)$, and $V(ec)$ are total volume of mitochondria, capillaries, and erythrocytes; $J(c)$ is total length of capillaries; Hct is hematocrit.

1:2.4:4.8, as seen in Table 2.1. In the two athletic species the relative muscle mass, M_m/M_b, is somewhat larger than in the goat but not in the same proportion, so that $\dot{V}_{O_2}max$ is not higher in athletes simply because of their larger active muscle mass. Instead we find that the muscle cells of the athletic species have increased the volume density of their mitochondria from 4% in the goat to 8.6% in the dog and 10.6% in the pronghorn. As a result the total mitochondrial volume in the entire locomotor muscle mass shows the proportion 1:3:4.6 in the three species and is thus about proportional to $\dot{V}_{O_2}max$. This suggests that the capacity for aerobic work is indeed determined by the amount of mitochondria in the muscle cells, and it appears that the mitochondria have invariant characteristics among these three species. When $\dot{V}_{O_2}max$ is divided into the mitochondrial volume, the rate of O_2 consumption per unit volume of mitochondria is similar in the three species, ranging from 4.6 in the dog to 5.9 ml $O_2 \cdot A$ min$^{-1} \cdot$ ml^{-1} in the pronghorn, a range of values that has been found to hold in many species, including the human.

From this comparison of three species with widely varying $\dot{V}_{O_2}$max we thus conclude that a higher capacity for oxidative phosphorylation is achieved simply by increasing the volume, of mitochondria in the muscle cells. The mitochondria are structurally and functionally the same: they operate, at $\dot{V}_{O_2}$max, at a similar rate of about 5 ml O_2 per minute and ml of mitochondria. Turning the argument around, we can also say that the muscle cells must incorporate 15 ml of mitochondria — or 525 m² of inner mitochondrial membrane — for every ml of oxygen they are consuming per second at $\dot{V}_{O_2}$max.

As an aside, let us carry this one step further and make a small calculation to see what this means in terms of the number of chemical reactions that occur on the enzymes of the respiratory chain. One ml of mitochondria contains, as we have seen, 35 m² of inner membrane. In a study that combined biochemical, physiological, and electron microscopic methods, Schwerzmann et al. (1989) estimated that 1 μm² of inner membrane carries about 9500 units of enzyme complex IV (cytochrome oxidase a-a$_3$, the last step in the respiratory chain where O_2 is consumed when the protons are oxidized to water (Figure 2.6). We therefore estimate that the entire inner membrane in 1 ml of mitochondria carries $3.3 \cdot 10^{17}$ such enzyme complexes. Now, 1 ml O_2 contains $2.7 \cdot 10^{19}$ oxygen molecules, as can be inferred from Avogadro's number. From this we estimate that the number of O_2 molecules that are consumed — or reduced to H_2O — in 1 ml mitochondria is on the order of $1.35 \cdot 10^{20}$ per minute or $2.25 \cdot 10^{18}$ per second. Dividing this into the number of enzyme units, we estimate that about 7 molecules of O_2 are being reduced at every complex each second when the muscle cells work at $\dot{V}_{O_2}$max, and this yields 42 molecules of ATP. It is interesting that biochemical studies have shown the maximal rate of enzyme activity, under optimal conditions in the test tube, to be about 15 molecules of O_2 per second per complex IV, so that the mitochondria in the working muscle cell appear to operate, on time average, at about half this theoretical maximal rate; and this is about the same in all three species.

Mitochondria thus are the containers of an enzyme system whose pathway begins with the Krebs cycle and terminates in the oxidase of complex IV. The relation of form and function is tightly regulated and conserved in that the carrier of the oxidase, the inner membrane, is made proportional to the space that contains the other enzymes, the matrix. A conservation of mitochondrial structure results, specifically of the inner membrane density, even if the need for oxidative phosphorylation changes by a factor of 5, as between the goat and the pronghorn. This is strong evidence in support of the hypothesis that structure is closely adjusted to the functional needs at the level of cell organization, and thus in support of symmorphosis.

We shall see in Chapter 7 that the constancy of mitochondrial structure is true for all mammalian species studied and that they all achieve a similar maximal rate of oxidation per unit mitochondrial volume. But there are interesting exceptions to this rule. For example, in hummingbird flight muscle mitochondria occupy about 50% of the cell space, leaving less than half for the actual force-generating machinery. But this does still not suffice to cover the extraordinarily high energy needs of the hummingbird's flight. In this instance it has been observed that the density of inner mitochondrial membrane within the mitochondrion becomes increased (Suarez 1998). These studies have led to an interesting conclusion: if $\dot{V}_{O_2}$max is related to the surface of the inner mitochondrial membrane, oxygen consumption per unit area of the inner mitochondrial membrane is the same in the hummingbird as in the mammalian muscles. Conversely, the inner membrane surface density is lower in the lizard, but again the ratio of $\dot{V}_{O_2}$max to the total inner membrane area is very close to the value found in mammals and birds (Conley et al. 1995).

We thus conclude that the aerobic capacity of the muscle cells is set by the mitochondrial inner membrane, presumably by the number of respiratory chain units it carries. If the cells need more aerobic ATP, they adapt to this functional requirement by building more mitochondrial inner membrane and, by that, more mitochondrial volume, since the ratio of inner membrane and matrix must be maintained. It appears furthermore that they build no more than is needed. To accommodate the contractile machinery and an energy-supply system within the muscle cell results in a rather crowded situation with no space to waste. The evidence that the design of muscle cells follows the rules of economy is compelling. Muscle cells are also adaptive in that they can change their

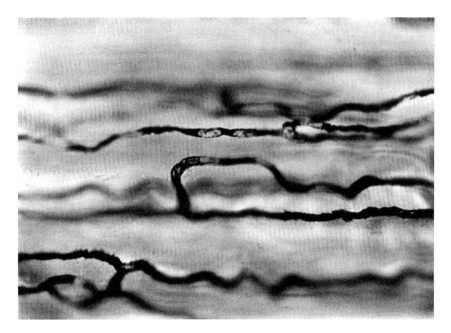

Figure 2.8 Skeletal muscle capillaries form networks that mainly course parallel to the muscle fibers.

quantity according to need. This all is strong support for the principle of symmorphosis in the relation of form and function at the level of the cell.

Are Muscle Capillaries Adjusted to Mitochondrial Oxygen Needs?

Mitochondrial oxygen needs are great: every minute 1 ml of mitochondria consumes about 5 ml of oxygen at $\dot{V}_{O_2}$max. The muscle cells contain a little bit of oxygen bound to myoglobin, but that is by no means sufficient. As a consequence, all the oxygen consumed by the mitochondria must be steadily and nearly instantaneously supplied from the blood. The blood flows past the muscle cells in a network of capillaries that run predominantly parallel to the muscle fibers (Figure 2.8). If we look at a cross-section through one of these capillaries (Figure 2.9), we

Cells and Tissues · · · 53

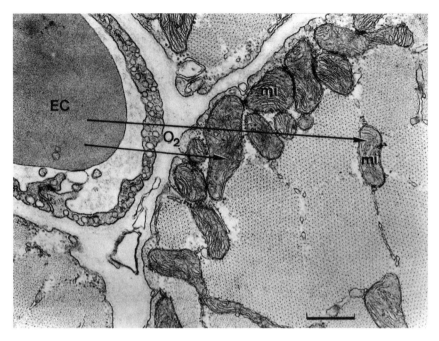

Figure 2.9 The peripheral pathway for oxygen begins in the capillary erythrocytes (EC) and ends in the mitochondria (mi), which are distributed throughout the muscle cell. (From Weibel 1984.) Scale marker: 0.5 μm.

see that the pathway for oxygen begins in the red blood cell, which contains oxygen at a high concentration, bound to hemoglobin. Oxygen diffuses through the blood plasma, through the endothelial cell and through the interstitium to reach the muscle cell, where it is transferred to the mitochondria along the myoglobin pathway, which facilitates oxygen diffusion.

The basic model describing the effect of capillary design on oxygen delivery to the muscle cell is one of the achievements of August Krogh. In his classic treatise *Anatomy and Physiology of Capillaries* (1922, 1929), he specifically dealt with muscle capillaries and designed a model that today is called the Krogh cylinder. Krogh conceived that the domain supplied with oxygen from a given capillary takes the form of a sleeve of muscle cells that contains the consumers, the mitochondria (Figure 2.10). Quantitatively, the radius of this cylinder is about half the distance between the capillaries; more precisely, its cross-sectional

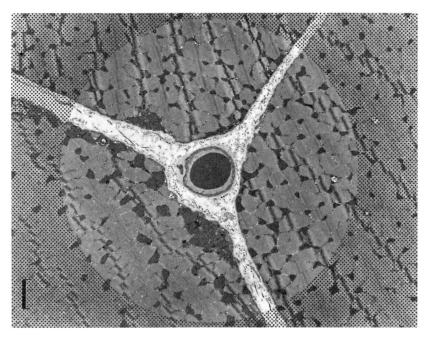

Figure 2.10 Electron micrograph of a cross-section of muscle cells with a capillary to show the concept of a Krogh cylinder centered on the axial capillary and containing mitochondria as O_2 consumers in the surrounding muscle cells. Scale marker: 2 μm.

area is inversely proportional to the density of the capillaries in muscle tissue measured by their length per unit volume of muscle. Since capillaries are found to have a fairly constant internal diameter of about 5 μm and the length of the capillary path is about 0.5 mm, the Krogh cylinder model establishes a rather precise relationship between capillary volume and muscle volume.

Krogh's primary objective was to estimate the change in O_2 delivery to the muscle cells as the blood traveled along the capillary from the arterial to the venous end. For that he had to estimate the rate of O_2 consumption of the surrounding muscle cells, and he hypothesized that the diameter of the cylinder should be smaller if the O_2 needs were greater. We know that the maximal O_2 needs are determined by the amount of mitochondria in the muscle cells. If the capillary supply to the muscle cells is designed to match the cells' O_2 needs, we predict that the diameter of the Krogh cylinder should be such as to keep the total volume of mitochondria in the tissue cylinder constant. This is indeed what we found when we studied different muscles in a small African antelope: in highly aerobic muscles with a mitochondrial density of 18%, the radius of the Krogh cylinder was about 13 μm, compared with 23 μm in less aerobic muscles with a mitochondrial volume density of 5%. As a result we calculated that, in either case, there were about 3 ml of mitochondria per ml of capillary blood, a relationship confirmed in other species.

These are average values obtained simply by dividing the total mitochondrial volume by the total volume of capillaries measured in the same muscle samples. But there are some interesting details in the design of the relation between capillaries and mitochondria that are worth mentioning. If one looks at Figures 2.9 and 2.10 one notices clearly that the mitochondria are not homogeneously dispersed throughout the muscle cell. Instead they are crowding toward the capillary, a feature that is found quite generally. Indeed, in nearly all muscle types of all species the concentration of mitochondria in the region near a capillary is about twice as high as it is farther away, in the center of the fiber. This indicates that fine structural design features are much more subtle than what we show in most quantitative data. Statistics sometimes—if not often—hide the most precious information, and attention must be paid to discover it.

To generalize from this simple model we predict that the capillary volume is proportional to the mitochondrial volume in the associated muscle cells and that it is, accordingly, also proportional to $\dot{V}_{O_2}$max. This prediction is based on the notion that no more capillaries should be built than are required to deliver oxygen, an idea that derives from the principle of symmorphosis.

What do we find if we study our adaptive series of goat, dog, and pronghorn? Table 2.1 shows that the capillary volume is indeed larger in the athletic species, and that it increases from the goat to the antelope in proportion to the ratio 1:1.8:2.7, thus less steeply than the ratio 1:2.4:4.8 we found for the volume of mitochondria and for $\dot{V}_{O_2}$max. Does this mean that the capillaries are not fully adjusted to the cells' need for oxygen supply? Yes and no. No, because our model was indeed too simple. Oxygen is brought into the capillary bound in erythrocytes, and we should in fact have predicted that the volume of capillary *erythrocytes* should be proportional to mitochondrial volume and $\dot{V}_{O_2}$max, rather than the volume of the capillary tube. Because blood is composed of blood cells and plasma, the volume of erythrocytes in the muscle capillaries is the product of the capillary tube volume and the volume density of erythrocytes in blood, which is called the hematocrit. In human blood sampled from a large vein the hematocrit is 0.45, which means that 45% of the blood volume is occupied by oxygen carrying red blood cells. Some animals, such as horses, can increase the red blood cell concentration during exercise, so that hemocrit is variable.

It is known, however, that the hematocrit is lower in capillaries than in large veins because the narrow capillary tube inhibits the passage of erythrocytes, which have about the same diameter. By how much the capillary hematocrit is lower is a matter of debate, but I will assume— though this is not firmly grounded—that it is proportional to measured venous hematocrit when the total red cell content varies among the species.

It turns out that the hematocrit is, indeed, variable. We have found that venous hematocrit is significantly larger by 50% in the two athletic species than in the goat (Table 2.1). The total capillary red cell volume corresponds to the ratio 1:3:4.1 in the three species, and thus is closely matched to the total mitochondrial volume in muscle cells and to $\dot{V}_{O_2}$max. Table 2.1 shows that the O_2 supply rate from red cells, estimated as $\dot{V}_{O_2}$max per erythrocyte volume, is very similar in the three species. This result is fully compatible with the hypothesis of symmorphosis. Oxygen supply to muscle cells depends on the design of two independent structures: the vascular tube system and the cells of the blood. The adjustment of the oxygen supply capacity of the capillaries is achieved by modulating two principal structural parameters, one characterizing the vasculature—the capillary volume—the other characterizing the blood—the hematocrit. It is indeed quite economic to split the adjustment between the two components.

Symmorphosis in the O_2 Pathway in Muscle

What can we conclude so far? Does the evidence we have collected support the hypothesis that form and function are closely related according to the principle of symmorphosis, or, in other words, that quantitative design parameters set the aerobic capacity of muscle?

In order to test this hypothesis I now introduce a test parameter of which we shall make more use later on, namely the ratio of the presumed critical design parameter, [X(.)], to the estimated functional capacity {Y(.)} which in our case is {$\dot{V}_{O_2}$max}. I now predict that, for symmorphosis to be supported, the ratio

$$[X(.)]/\{Y(.)\} = [X(.)]/\{\dot{V}_{O_2}\text{max}\}$$

is invariant when $\dot{V}_{O_2}$max varies.

With respect to the mitochondria the ultimate design parameter is the inner mitochondrial membrane area S(imi), which determines the number of respiratory chain units and is the product of total mitochondrial volume V(mi) multiplied by the inner membrane surface density in the mitochondria. Since we found the latter parameter to be invariant among the mammalian species studied, and since we found mitochondrial volume to be proportional to $\dot{V}_{O_2}$max, we conclude that the two ratios

$$[V(mi)]/\{\dot{V}_{O_2}\text{max}\} \text{ and}$$

$$[S(imi)]/\{\dot{V}_{O_2}\text{max}\}$$

are invariant, thus supporting the hypothesis at the level of the muscle cells.

The situation is not as simple for the capillaries. The first prediction that the capillary volume V(c) should be proportional to $\dot{V}_{O_2}$max is not supported by the evidence. The model for oxygen delivery to the muscle cells involves two structures, the vasculature and the blood, and we find that both structures contribute to the adjustment to higher oxygen delivery in the athletic species. In fact, both contribute equally to adjustment when the athletic species make more capillaries and pack more erythrocytes into their blood. As a result, the critical design parameter is the product of two independent structural variables, capillary volume and the volume density of red cells or hematocrit V_V(ec,c), which forms an invariant ratio with functional capacity:

$$[V(c) \cdot V_V(ec,c)]/\{\dot{V}_{O_2}\text{max}\}$$

This is a highly interesting result in light of our basic hypothesis that economy is one of the fundamental strategies in the design of an organ-

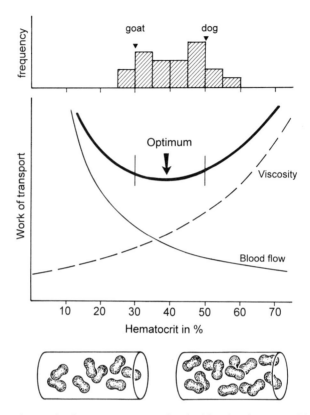

Figure 2.11 The work of oxygen transport by the blood is determined by the blood flow and by the resistance due to viscosity. These two factors are affected by hematocrit in opposite ways, resulting in a range of hematocrits that is not far from optimal. The top graph shows that hematocrit measured in different mammalian species falls into this range of near-optimal hematocrits.

ism. Of course, it is conceivable that the muscle could adjust its capillary bed to be 2.5 times larger in the dog and 5 times larger in the pronghorn. What counts, however, is the number of oxygen-carrying erythrocytes on the surface of muscle cells, and it makes no difference whether the adjustment is made by enlarging the capillary bed or by packing more erythrocytes into it; the number of red cells required is the same. Thus it definitely appears economic to make part of the adjustment for higher oxygen-delivery capacity by increasing the hematocrit and thereby saving on the need for additional capillaries.

There is a limit, however, to how far the hematocrit can be increased. With increasing hematocrit, blood viscosity increases, which causes blood flow resistance to rise. There, in fact, appears to be an optimal hematocrit (Figure 2.11). Increasing the concentration of erythrocytes in the blood has both a positive and a negative effect: oxygen transport becomes cheaper because less blood must be pumped to the periphery to transport one unit of oxygen, but conversely the heart must do more work as resistance increases due to higher viscosity. The total work to be done to transport a unit of oxygen is the sum of these two functions, shown as a heavy line in Figure 2.11. We see that this function goes through a minimum at a hematocrit of about 40%. The top graph shows the range of hematocrits that are found in mammals, with the values found in the dog and the goat indicated. It is interesting that these values lie at the edge of an optimum range between 30% and 50%, where variations in hematocrit, in fact, have a small effect on the work of transport. This is further evidence that the design of tissue structures tends toward economic, perhaps even optimal, solutions.

3: Muscle: Supplying Fuel and Oxygen to Mitochondria

The model of the pathway leading to oxidative phosphorylation shown in Box 2.1 makes it evident that oxidative phosphorylation can only proceed if fuel is supplied to the mitochondria at a rate matched to oxygen consumption. The fuels are mostly of two kinds: carbohydrates, chiefly represented by the 6-carbon sugar glucose, and free fatty acids with their hydrocarbon chains some 16 carbons long. Proteins can also be used as fuel, but their contribution is insignificant for fueling exercise and will not be considered here.

The oxidative metabolism of glucose and fatty acids is a tightly controlled process that uses the Krebs cycle as the final common pathway (Box 2.1). From the Krebs cycle onward there is no difference between fuels derived from carbohydrates or fatty acids. The processes that generate the acetyl-CoA needed to initiate the cycle are very different for glucose and for fatty acids, however, and this affects the stoichiometry of the total process. The catabolism of glucose begins in the cytoplasm with glycolysis, which splits the glucose molecules into two 3-carbon pyruvates and generates two ATP and NADH. Glycolysis is an anaerobic process not dependent on oxygen; some oxidized NAD must be available, though, which can be provided by the mitochondria but can also be obtained from the NADH generated by reducing pyruvate to lactic acid. Lactic acid then becomes the end product of anaerobic glycolysis, to be released into the blood (Figure 2.2). Under aerobic conditions, that is, when there is enough O_2, pyruvate is shuttled into the Krebs cycle, and the overall product of oxidation is 36 mol ATP for every mol of glucose metabolized—thus 18 times more than by glycolysis alone—and this consumes 6 mol O_2 and produces 6 mol CO_2.

The metabolism of fatty acids is in a way simpler, but it is only possible under aerobic conditions. A process called β-oxidation clips off 2-carbon units from the long hydrocarbon chain in the process of making acetyl-CoA for entry into the Krebs cycle (Figure B2.1). The fatty acids vary in chain length, but for a typical fatty acid the stoichiometry says that 130 mol ATP are generated by the oxidation of 1 mol palmitic acid, which consumes 23 mol O_2 and produces 16 mol CO_2.

Putting all this together, we can partition the O_2 flow rate, expressed in molar units,* $\dot{M}_{O_2}$, into that part derived from carbohydrate, $\dot{M}_{O_2}^{CHO}$, and that derived from fatty acids, $\dot{M}_{O_2}^{FFA}$, as described in detail in Box 3.1.

It is evident that oxidative phosphorylation can only proceed if the supply of both oxygen and fuels is matched to the cells' energetic needs. We have so far argued under the assumption that oxygen supply is the critical process. But is this justified? Is the supply of fuels not just as critical? In particular, we must ask whether the design of capillaries is indeed matched to the needs for oxygen supply, as we have surmised in the preceding chapter, or whether the provision of substrates is not just as important.

Differences between Oxygen and Fuel Supply

Before proceeding further I must point out the important differences between oxygen and substrates:

1. In the blood, oxygen is transported in erythrocytes whereas fuels are contained in the plasma. Glucose is water-soluble and can therefore be carried in solution; fatty acids are highly hydrophobic and must be solubilized by binding to albumin.

2. The transfer of oxygen from the capillary into the cell occurs purely by diffusion since oxygen is soluble in both aqueous and lipid media and can thus easily traverse lipid membranes. This is very different for the substrates. The water-soluble glucose can diffuse freely through the aqueous spaces plasma, interstitium, and cytoplasm, but it cannot readily cross the lipid barriers of cell membranes in the endothelium and sarcolemma; specific glucose transporter proteins called

---

* In this chapter I express the O_2 flow rate in molar units rather than in volume so as to allow correlation with substrate flux. Conversion: 1 mol O_2 = 22.39 1 O_2 (at 0°C, 1 atm) or 1 ml O_2 = 44.6 µmol O_2.

Box 3.1 The stoichiometry of fuel supply to mitochondria in muscle

The four pathways for fuel supply to mitochondria are shown in Figure B2.1 and also in Figure 3.2. The carbohydrate glucose (CHO) and the free fatty acids (FFA) each have two ways of reaching the mitochondria, either directly from the intravascular plasma pools (iv) or via intracellular stores (ic) deposited during periods of low activity. Their molar flux rates per unit of time are symbolized as follows:

1. $\dot{M}_{CHO}(iv)$—glucose flux from plasma in capillary
2. $\dot{M}_{CHO}(ic)$—glucose flux from intracellular glycogen stores
3. $\dot{M}_{FFA}(iv)$—free fatty acid flux from plasma in capillary
4. $\dot{M}_{FFA}(ic)$—free fatty acid flux from intracellular lipid stores

The flux rates of these fuels into the mitochondria are

$$\dot{M}_{CHO}(mi) = \dot{M}_{CHO}(iv) + \dot{M}_{CHO}(ic)$$
$$\dot{M}_{FFA}(mi) = \dot{M}_{FFA}(iv) + \dot{M}_{FFA}(ic)$$

In Box 2.1 it was shown that the oxidation of these fuels in the mitochondria begins with the introduction of acetyl-CoA into the Krebs cycle (Figure B2.1) and then leads through the common pathway to the terminal oxidase in the respiratory chain. Oxidation of 1 mol glucose by this pathway requires 6 mol O_2 and yields 6 mol CO_2, whereas the oxidation of 1 mol fatty acid (palmitic acid) consumes 23 mol O_2 and yields 16 mol CO_2. The stoichiometry of total oxidation of these two substrates is therefore

$$\dot{M}_{O_2} = \dot{M}_{O_2}^{CHO} + \dot{M}_{O_2}^{FFA}$$
$$= 6\,\dot{M}_{CHO} + 23\,\dot{M}_{FFA}$$
$$\dot{M}_{CO_2} = 6\,\dot{M}_{CHO} + 16\,\dot{M}_{FFA}$$

The ratio $\dot{M}_{CO_2}/\dot{M}_{O_2}$ is called the respiratory exchange ratio (RER) and is different for the two fuels because of the difference in their CO_2 yield: $6/6 = 1$ for glucose and $16/23 = 0.71$ for fatty acids. If both glucose and fatty acids are oxidized, the measured RER is between 0.71 and 1.0, and the fraction of fatty acid burned is obtained by

$$\dot{M}_{O_2}^{FFA} = [\dot{M}_{O_2} \cdot (1 - RER)]/0.29$$

and $\dot{M}_{O_2}^{CHO} = \dot{M}_{O_2} - \dot{M}_{O_2}^{FFA}$

Thus the relative contributions of the two substrates to total oxidation can be estimated by measuring O_2 consumption and CO_2 production in the breathing air, a procedure called indirect calorimetry. For example, if we measure RER to be 0.9 we calculate that 34% of the fuel was fatty acids and 66% glucose.

· · · ·

GLUT-1 and GLUT-4 are built into the sarcolemma in order to carry the glucose molecules from the interstitial space into the cytoplasm. Similar carriers are also found in the endothelial membranes, but there the transfer of glucose from the plasma to the interstitial space can proceed by diffusion through the intercellular junctions, as we shall see in some detail below. For fatty acids the situation is reversed: they can traverse membranes, but they cannot freely move through aqueous spaces. Accordingly, they must be solubilized by binding to albumin both in the plasma and in the interstitial space, and the cytoplasm is provided with a special fatty acid binding protein for their transfer from the cell surface to the lipid stores, the triglyceride droplets near the mitochondria. In a way, albumin and fatty acid binding protein function as carriers for fatty acids through the aqueous spaces.

3. Perhaps the most important difference between oxygen and substrates is the fact that neither the cells nor the body as a whole has the capacity to store significant amounts of oxygen. In contrast, abundant stores are provided both within the muscle cells and in specialized storage organs for carbohydrates as well as for lipids. The storage form of glucose is glycogen, a glucose polymer similar to starch which is found in the form of small granules in the cytosol of most cells—including muscle cells, where they lie close to mitochondria as well as within myofibrils (Figure 3.1a); larger packets of glycogen granules are found in hepatocytes of the liver, which serves as the main carbohydrate storage organ. Fatty acids are stored as water-insoluble triglycerides, which form small droplets in the cytoplasm of muscle cells in close relation to mitochondria (Figure 3.1b); the major storage organ of triglycerides is

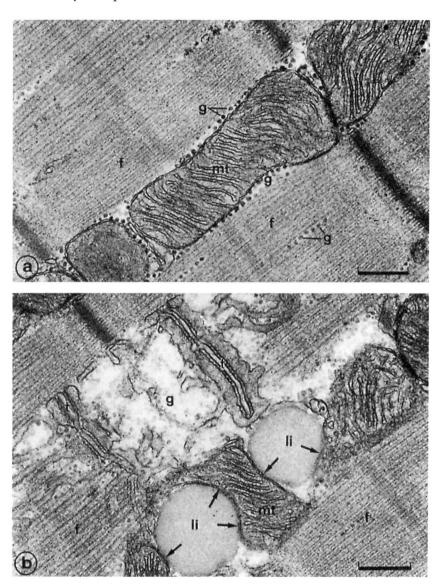

Figure 3.1 Electron micrographs of muscle cell showing the distribution of intracellular fuel stores: (a) glycogen granules (g) are mostly found at the periphery of mitochondria (mt) but also inside myofibrils (f). (b) Lipid droplets (li) are tightly associated with mitochondria (arrows). Note t-tubules (t), which are in close contact with the tubules of sarcoplasmic reticulum, the organelles that control Ca^{++} flux. (From Vock et al. 1996a.)

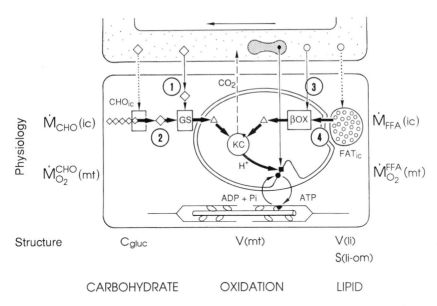

Figure 3.2 Four pathways of fuel supply for oxidation in mitochondria, two for glucose (1, 2) and two for fatty acids (3, 4), whereby (1) and (3) supply substrates for glycolysis (GS) directly from the capillary, (2) and (4) from intracellular stores. (Modified after Vock et al. 1996a.)

fat tissue, where adipocytes typically contain one large triglyceride droplet. These forms of fuel stores are important design features of the fuel supply system both in muscle cells and in the storage organs.

The model describing the pathways for oxygen and substrates from the capillary to the mitochondria, shown in Figure 3.2, must account for these differences. All pathways converge at the respiratory chain in the inner mitochondrial membrane, which oxygen reaches directly by diffusing from the red blood cells in the capillaries. The pathways for the two substrates are more complicated. They converge at the Krebs cycle for the common final pathway, as noted above (see also Box 2.1). In the glucose pathway, acetyl-CoA is derived from glycolysis, requiring a constant influx of glucose provided by two different routes: (1) glucose supplied from the intravascular glucose pool in plasma can go directly into glycolysis; (2) glucose can be provided through a broken pathway in which plasma glucose is first polymerized into glycogen as storage carbohydrate, which is subsequently broken down to provide phosphorylated glucose for glycolysis.

Similar conditions prevail for the fatty acid pathway (Figure 3.2): some fatty acid is directly provided from plasma to β-oxidation within the mitochondria (3); the major pathway, however, is through triglyceride stores (4) where fatty acids brought in from the plasma are first esterified to triglyceride (triacylglycerol), which are deposited into lipid droplets. These droplets are directly associated with mitochondria (Figure 3.1b). In the region of contact, enzymatic lipolysis liberates fatty acids, which are transferred into the mitochondria through a special carrier system, called the carnitine shuttle, in order to be introduced into β-oxidation.

The important feature of these substrate supply pathways is that, for both glucose and fatty acids, the substrates can be drawn either directly from the blood plasma or from intracellular stores, a supply route which is indirect because the stores are ultimately stocked up from the plasma sources. Thus we can formally state that, for both carbohydrates and fatty acids, the flux of substrates into the mitochondria (mi) during muscle work is the sum of substrate flux from intravascular sources (iv) and from intracellular stores (ic), as described in detail in Box 3.1.

How is the substrate supply partitioned between carbohydrate and fatty acid, on the one hand, and between direct (iv) and indirect (ic) pathways, on the other hand? We must also ask whether this partitioning is associated with specific design features and whether these can be interpreted in terms of the hypothesis of symmorphosis. This is, of course, a much more complex situation than that discussed in Chapter 2. We must, indeed, ask whether we may have been misled by considering a too simple case when limiting ourselves only to O_2 supply. The critical question is whether the capillaries are designed for O_2 supply, or whether they are rather designed for fuel supply—or as a compromise between the requirements of the two.

Variations in Fuel Supply to Mitochondria in Working Muscle Cells

We will approach this question by again comparing dogs and goats, two species that differ greatly in their need for oxidative metabolism. As already noted, the athletic dog has a capacity for oxidative metabolism that is 2.5 times greater than that of the goat when expressed per kg body mass (Table 2.1); if we express $\dot{V}_{O_2}$max relative to the muscle mass, which is somewhat larger in the dog than in the goat, it is still greater in the athletic species by a factor of 1.6. We therefore expect that total substrate oxidation is proportionally greater in the muscles of the dog than in those of the goat, but the question must be asked whether fatty acid and glucose contribute equally in both species and to what extent the substrates are drawn from intravascular sources or from intracellular stores. If one or the other of these pathways has critical design properties, we would expect to find differences between the two species. This would then allow us to ask how these different flux rates for substrates and oxygen relate to the relevant structures, with the final question being whether the notion of symmorphosis can be applied to connected pathways, that is, to pathways that partly use the same structures and have the same ultimate goal: the mitochondrion. Clearly this study is much more involved than what we considered in Chapter 2, as it combines physiological studies with biochemical analysis and finally morphometry. This is a study that can only be performed by a team of well-qualified specialists in different fields, specialists whose understanding of biology is broad enough to allow them to communicate across the boundaries of the various disciplines. I have had the privilege of working with such a group.

Let us now ask three questions directed to clarifying the relation between form and function in this complex system at the level of the muscle cells and their associated capillaries:

1. How is consumption of the two substrates upregulated in exercise?
2. How much of these substrates is directly supplied from the capillaries?
3. How is this related to morphometric properties of capillary and muscle cell design? Specifically, can we identify design features that limit substrate flux?

These questions may be answered by studying the consumption of fuels as the animals increase the relative intensity of their exercise and their concomitant O_2 consumption as they run on a treadmill.

Partitioning of Fuel Consumption between Glucose and Fatty Acids

In Figure 3.3 the oxidation rates of carbohydrate and fatty acids is plotted as a function of increasing exercise intensity, given as the percentage of maximal oxygen consumption. As expected, total oxidation increases linearly with exercise intensity, and it is, at every exercise intensity, proportionately higher in dogs than in goats.

The question of how much of this total oxidation is contributed by carbohydrate and fat oxidation respectively can be answered by indirect calorimetry, outlined in Box 3.1. It exploits the fact that fatty acid oxidation yields relatively less CO_2, namely 16 mol for 23 mol O_2 consumed, whereas the oxidation of glucose yields the same amount of CO_2 as O_2 is consumed. The respiratory exchange ratio RER = $\dot{M}_{CO_2}/\dot{M}_{O_2}$ is therefore 1 when glucose is oxidized, and 0.71 if fatty acids are oxidized. All the CO_2 produced is eliminated through the lungs, so that we can measure the rate of CO_2 production from exhaled air just as we can measure O_2 consumption from the depletion of oxygen in air during respiration. We can then calculate RER by the equation given in Box 3.1. What is required, however, is that the animal reaches a steady state in its fuel consumption, and this cannot be achieved at $\dot{V}_{O_2}$max; the highest exercise intensity where this is possible is 85% of $\dot{V}_{O_2}$max, but that is fairly close to the limit of performance.

In this way the partitioning of carbohydrate and fatty acids was calculated at different exercise intensities in dogs and goats, as illustrated in Figure 3.3. The most interesting finding is that the maximal rate of fatty acid oxidation was already reached at 40% exercise intensity in both the dogs and the goats, so that all the additional energy required as exercise intensity increased was obtained through combustion of glucose. This is often referred to as the "cross-over concept" since the *relative* fuel "preference" switches from fatty acid to glucose as exercise intensity increases, but it is essentially due to the fact that the fatty acid supply cannot be increased any further. We conclude from this that the two substrates behave very differently in fueling muscle exercise, but that the pattern is the same in both dogs and goats. Furthermore, it is also evident that dogs use more of both substrates than goats.

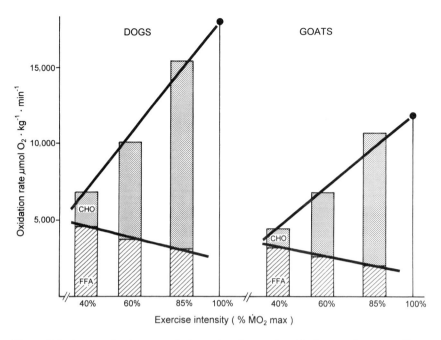

Figure 3.3 As exercise increases in intensity, the total oxidation rate increases linearly whereas the relative rate of lipid oxidation falls. The increase in work rate is therefore fueled almost exclusively by glucose. (From Roberts et al. 1996.)

Estimating the Capillary Supply of Substrates

How can we estimate the fraction of the glucose burned in the mitochondria that is derived directly from the circulating pool of glucose in plasma (route 1 in Figure 3.2)? This can be done by infusing into the blood glucose that has been radioactively labeled by replacing one hydrogen atom with tritium. By taking blood samples every few minutes during the run the rate of disappearance of glucose into the muscle cells can be estimated. A similar procedure can be used for fatty acids by infusing palmitate labeled with a ^{14}C atom. The appearance of $^{14}CO_2$ in plasma can be used as an estimate of palmitate oxidation from circulatory sources.

The question to be answered was whether substrate flux from the capillaries increases in parallel with oxygen consumption as blood per-

70 · · · Symmorphosis

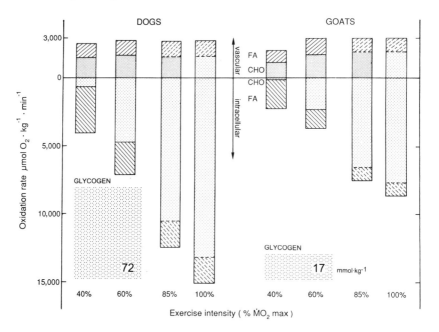

Figure 3.4 In both dogs and goats the vascular supply of glucose and fatty acid is the same and does not increase as exercise intensity increases. The higher rate of oxidation in the dog is achieved by drawing to a greater extent on the intracellular stores of glycogen, which are larger in the dog than in the goat (stippled rectangles). (Modified after Weibel et al. 1996.)

fusion of the muscle is upregulated with increasing exercise. The results are shown in Figure 3.4 and reveal that this is clearly not the case. For both carbohydrate and fatty acids, the supply rate from circulating vascular sources reaches its maximum at 60% of exercise intensity and does not increase further as exercise intensity increases. A further interesting observation: the vascular supply rate for both fuels is not higher in the dog than in the goat. Because of the greater total substrate oxidation rate, however, the vascular substrate supply is a smaller fraction of total substrate combustion in the dog. Also, this fraction decreases as exercise intensity increases, in both species.

This leads to four conclusions:

1. The supply of fuels from the capillaries is strictly limited at the same absolute level in both dogs and goats.

2. The supply of fatty acids from intracellular lipid stores is higher in the dog than in the goat but it does not increase with increasing exercise intensity.
3. The increase in fuel supply to the mitochondria with increasing exercise intensity is covered from intracellular stores of glycogen in both species.
4. The higher fuel needs of the dog are covered by drawing more on the intracellular glycogen stores and not by supplying more from the capillaries.

Simply on the basis of these physiological findings we must refute the hypothesis that microvascular design is sufficient for both oxygen and substrate supply: it suffices for oxygen, but not for substrates. The question arises whether this limitation is related to design properties of the supply pathway from the capillary plasma to the mitochondria. I will limit this discussion to glucose, since only glucose oxidation is increased as oxidative phosphorylation is increased in exercise, and dogs and goats differ mostly in terms of the rate of glucose oxidation.

Revising the Model for Capillary Oxygen and Substrate Supply

The electron micrograph of Figure 3.5 shows that the flux of glucose from the plasma into the muscle cell meets with two resistances in the form of plasma membranes, which are impermeable to nonlipophilic solutes such as glucose. The first such barrier is constituted by the endothelial cells that separate the interstitial space from plasma. However, it turns out that this is not really a strong barrier to glucose flux, because the junctions between endothelial cells are relatively leaky and provide a system of pores for the diffusion of glucose from the plasma to the interstitial space. Such junctions, as shown in Figure 3.6a, are the seams between neighboring endothelial cells, narrow clefts that form a network of lines on the capillary surface.

Much more significant is the second barrier, the sarcolemma, an uninterrupted lipid membrane that does not allow the passage of aqueous solutes (Figure 3.6b). However, the sarcolemma is made permeable for glucose by a set of glucose transporters, specific protein molecules that

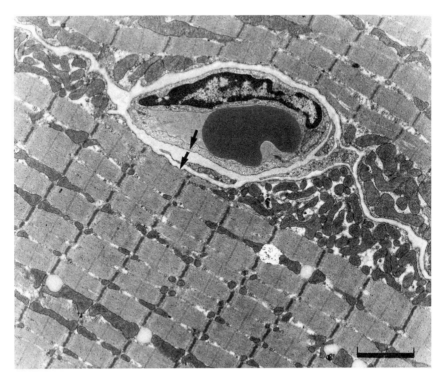

Figure 3.5 The capillary delivers both oxygen and substrates to the muscle cell. The oxygen derives from erythrocytes, the substrates from plasma. The arrows mark the two barriers for substrate flux, endothelium, and sarcolemma, respectively. Scale marker: 2 μm.

assemble in the membrane to form pores that facilitate glucose entry into the muscle cell. There are primarily two types of these transporters: GLUT-1 ensures baseline glucose permeability of the sarcolemma, and GLUT-4 is rapidly built into the sarcolemma from an intracellular pool in small membrane vesicles when glucose needs increase, such as in exercise. It appears, however, that the membrane becomes loaded with all the GLUT transporters it can accommodate at relatively low exercise intensities, so that at the high exercise levels considered here all GLUT reserves have been recruited and the highest permeability for glucose has been achieved.

On the basis of these observations I now postulate that the maximal flux of glucose into the muscle cells is determined by the conductivity of the sarcolemma, which is essentially determined by the sarcolemmal surface area together with the density of GLUT transporters, that is, the number of transporter molecules built into the unit area of the membrane. Unfortunately we know nothing about the density of GLUT transporters in dog and goat sarcolemma since most studies have been done in either rodents or humans, but there is no indication that there should be significant or even large interspecies differences.

We measured, in our dogs and goats, the sarcolemmal surface and found it to be identical in the two species (Figure 3.7). At first sight this appears astonishing, but it may not be. The main functional importance of the sarcolemma is the coupling of excitation and contraction of the muscle cells, as explained in Chapter 2, and we would not expect any difference between dog and goat in that respect. If we now look at the flux rate into the muscle cell of glucose from the intravascular pool in blood plasma, we note from Figure 3.7 that it is the same in dogs and goats, and that it does not increase between 60% and 85% maximal exercise intensity (Figure 3.4). As a result, the flux rate of glucose from vascular sources across the unit surface of the sarcolemma is the same in dogs and goats, and it does not increase with additional demand for glucose as the need for oxidative metabolism increases with exercise intensity.

These data suggest that the limitation of glucose uptake into the muscle cell is due to the limited permeability of the sarcolemma, which cannot be upregulated beyond that achieved at a 60% exercise intensity in both dogs and goats, and they also lend support to the assumption that the density and activity of the GLUT transporters is about the same in dogs and goats.

In contrast, the capillary surface S(c) is found to be significantly larger by a factor of 1.5 in the dog than in the goat (Figure 3.7); we also know that the length of intercellular junction lines that contain the pores through which glucose crosses the endothelium is proportional to S(c). So combine these observations: the flux of glucose from the plasma is the same in both species and the total surface of the endothelial pores is larger in the dog; accordingly the flux of glucose across the unit surface of endothelial barrier is 40% *smaller* in the dog than in the goat despite the dog's *greater* fuel needs. This paradoxical finding suggests that the endothelium is not a significant barrier to glucose flux from the plasma into the muscle cell.

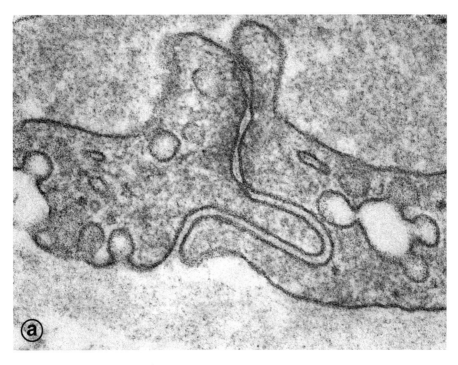

Figure 3.6 Barriers opposing substrate supply from capillary to muscle cell cytoplasm. (a) Capillary endothelial cell membranes are barriers to solute transfer, but the intercellular junctions are leaky clefts through which glucose can freely diffuse. In contrast, the sarcolemma (b) forms a single tight membrane barrier made permeable for glucose by the incorporation of specific glucose carrier proteins. ([a] from Vock et al. 1996b.)

Fuel Supply from Capillaries versus Intracellular Stores

These findings serve to refute the hypothesis that the design of the muscle microvasculature is adjusted to the supply of both oxygen and fuels. We very clearly found that the supply of substrates from circulating sources in the capillary is limited to a relatively low level such that, at high exercise levels, only about 20% of the fuels required by the mitochondria are provided by the capillaries. For glucose the evidence is strong that the barrier limiting the influx of substrate is the sarcolemma, whereas for fatty acids the precise location of the major resistance is as yet undetermined. The design parameters of capillaries bear

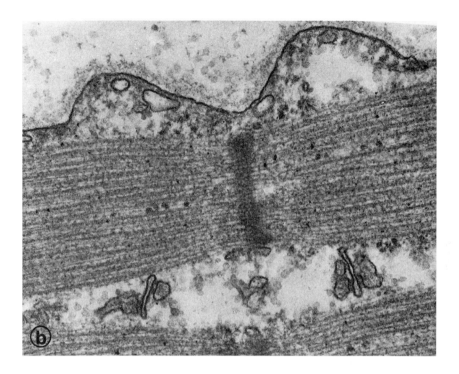

no quantitative relationship to the substrate flux rate; we conclude that the capillaries are indeed designed for oxygen supply to the muscle cells and their mitochondria, as found in Chapter 2. They seem to be adequate to supply glucose at the low rate permitted by the sarcolemma, but certainly not to cover the entire substrate needs of the working muscle cells.

The limitation of fuel supply from vascular sources is no disadvantage to the muscle cells, however, because they contain an ample store of these substrates in the form of glycogen granules and lipoprotein droplets. If we again refer to Figure 3.7 we notice that the supply of glucose from intracellular glycogen stores is about 1.6 times higher in the dog than in the goat at high exercise intensities, and that it is indeed several times greater than the supply from vascular sources, with the exception of the low exercise intensity of 40% (Figure 3.4). Dogs thus draw much more glucose from their intracellular glycogen stores, and it is therefore interesting that the muscle cells of the dog contain glycogen

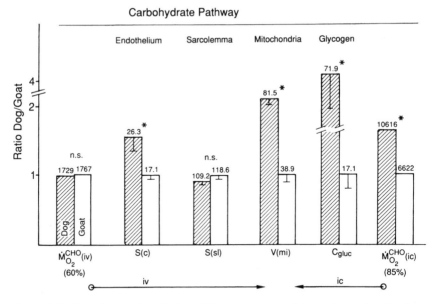

Figure 3.7 Quantitative description of the parameters of the pathway for the glucose supply to the mitochondria measured by their volume [V(mi)]: from left, the capillary supply [$\dot{M}_{O_2}^{CHO}$(iv)] across the endothelial [S(c)] and sarcolemmal [S(sl)] surfaces; from right, the supply of carbohydrate from intracellular stores [$\dot{M}_{O_2}^{CHO}$(ic)] recruited from the glycogen pool [C_{gluc}]. An asterisk indicates that the difference is significant; n.s., nonsignificant. (Modified after Weibel et al. 1996.)

stores which are about four times greater than those in the goat. This means that the dog, although it needs to draw almost twice as much glucose from its glycogen stores, can maintain this for a longer time, twice as long as the goat. The dog also has lipid stores which are more than two times greater than those in the goat so that, on the whole, the dog has twice the substrate reserve as the goat. It is thus endowed with the capacity for longer runs, a very important feature for a notorious endurance athlete.

There may be another reason why the glycogen stores must be larger than what is required to fuel high rates of oxidative metabolism: the relatively inefficient use of glucose in anaerobic glycolysis, which produces 18 times less ATP than oxidation in the mitochondria. Yet animals do have an occasional need for rapid supplementary ATP synthesis through anaerobic glycolysis when sudden sprints must be performed during a run and the speed exceeds that of $\dot{V}_{O_2}$ max. Such sprints consume very large quantities of glycogen since the end product is lactate. Seemingly excessive glycogen stores are therefore also an emergency reserve to provide for such occasions.

For muscle cells to depend primarily on intracellular stores of substrates to fuel high rates of muscle work appears problematic because it introduces the risk that the stores will become exhausted, but, in fact, from other points of view it makes a great deal of sense. The risk of exhaustion is compensated by the benefit of drawing on substrate pools that are located in the immediate vicinity of the mitochondria. This is very efficient in times of high demand, such as in exercise, because there are no structures interposed that would inhibit the recruitment of the substrates. Perhaps the best evidence of this is provided by the lipid droplets, which are found to be very closely associated with the mitochondrial outer membranes (Figure 3.1b). This association strongly suggests a direct recruitment of fatty acids from the intense contact surface between the outer mitochondrial membrane and the triglyceride droplet. It is at this surface that a lipase clips the fatty acids off the glycerol backbone of the triglycerides and transfers them to the carnitine shuttle, which brings them across the mitochondrial membranes to the site of β-oxidation (Box 2.1). A shorter pathway cannot be imagined. The situation is similar for the glucose pathway, where pyruvate generated by glycolysis in the cytoplasm is converted into acetyl-CoA at the outer mitochondrial membrane for entry into the matrix.

Furthermore, as we shall see in Chapter 8, the loading of intracellular fuel stores, ultimately from the gut, need not take place during the periods of high fuel demand but can take place at low rates during periods of rest, making use of the possibility to deposit these fuels in condensed form in the cell.

Conclusions on Form and Function in Muscle Cells and Tissue

In Chapters 2 and 3 we have looked in some detail at the quantitative design features of muscle cells and their associated capillaries in relation to their varying functional capacity. What have we found?

1. We noted that mitochondria are the sole site for oxidative phosphorylation in the muscle cells, and we have found that their volume is matched to the aerobic capacity of muscle cells, thus essentially determining V_{O_2}max.

2. We have also noted that a steady supply of oxygen to the mitochondria is required to come from the circulating blood in the capillaries, since the cells are unable to store oxygen in significant quantities. We have found that the volume of capillary erythrocytes determines the supply capacity of capillaries for oxygen and that this parameter is adjusted to the O_2 demand set by the mitochondria.

3. We have found that the supply of substrates from circulating pools in the capillaries is limited not so much by the design properties of the capillaries as by the transfer capacity of the sarcolemma and that this limited conductance for substrates, at least for glucose, appears to be invariant between athletic and sedentary species.

4. We have found that the supply of substrates to the mitochondria is regulated to needs by drawing on intracellular stores, and furthermore that the size of these stores is quantitatively adjusted to the demand, that is, to the endurance aerobic capacity of the cells.

So, finally, what can we conclude with respect to the simple question of whether cells and tissues are designed economically in conformity with the hypothesis of symmorphosis? The overall answer is not as simple as it was with respect to the relation between mitochondria and oxygen consumption. But I believe that we have collected considerable evidence to prove that economic design is an important principle. It appears that the cells make just the quantity of mitochondria they need to provide energy for muscle work at high rates. This is a simple case, one with a clear and unambiguous relation between one structure and one function: oxidative phosphorylation occurs exclusively in the mitochondria. Since the internal structure of mitochondria appears fixed and tightly adjusted to the functional processes at the molecular level, the only way to increase the capacity for oxidative phosphorylation is to increase the volume of the mitochondrial complement of the muscle cells. The fact that this is strictly proportional to the varying capacities for aerobic work between different species is strong evidence that the cells build no more mitochondria than they need.

The other processes studied are not as simple. Fuel supply to the mitochondria can take four different routes, which are affected by the design properties of different structures. The cells make use of all these options but favor, when the need for substrates is high, those that are most efficient: the recruitment of intracellular stores that are stocked up in periods of low activity. Because of this—even preferred—option of taking fuels from local stores, the design of capillaries is not critical for fuel supply, so oxygen supply to the mitochondria becomes the determinant factor of capillary design. The capacity of microvasculature for O_2 supply can be adjusted in two ways: by increasing capillary volume and/or by increasing the concentration of erythrocytes in the blood. We have seen that both options are used to provide a capacity for O_2 supply commensurate to the O_2 needs of the mitochondria. The burden of adjusting the structural design to higher functional demands is thus shared between the two structures involved, capillary vessels and the blood, and this, I believe, is one of the strongest pieces of evidence in favor of an integrated economic design strategy, and thus for symmorphosis.

4: Organ Design: Building the Lung as a Gas Exchanger

In this and the following two chapters I discuss how form and function are related at the level of an organ. Specifically, I ask whether the organ's complex design matches the functional demands imposed on it by the organism. In that sense I wish to explore whether the principles of economic design, adaptation, and integration of function, introduced in Chapter 1 in general terms and discussed in Chapters 2 and 3 with respect to cells and tissues, also prevail at the level of organ structure, in particular in the structure of the lung.

I choose the lung for several reasons. First of all, the lung's main function is well defined. In mammals it is the only organ that takes up oxygen from the air and transfers it to the blood at the rate needed by the cells to support oxidative metabolism. At the same time it also serves to eliminate the carbon dioxide (CO_2) that results from oxidative metabolism and needs to be discharged into air — where it is, by the way, recycled into oxygen and carbohydrates in photosynthesis by the plants, driven by the rich energy of the sun. This chief function provides the second reason for choosing the lung as an example, namely its intimate relation to the main case study of this book, the design of the system for energy supply to the organism. As we have seen in the preceding chapters, oxygen must be supplied to the cells at the rate imposed by the need for oxidative phosphorylation by the mitochondria. And it must be supplied by the blood instantaneously as the muscle cells work and need energy. Since the body cannot store oxygen in any significant amount, O_2 must be taken up from the air and loaded onto the blood nearly instantaneously as well. In this process the lung plays the critical role as the only gas exchanger between air and blood.

To achieve this, the lung is in a way a spectacularly designed organ, one that must bring blood into very intimate contact with the outside air while still keeping the two media separate. Its main features are a complex system of airways and blood vessels that access an extraordinarily large internal surface area which in humans approaches the size of a tennis court. With this lavish outfit it is indeed hard to conceive that the principle of economic design should prevail in the lung, unless we can show that this large a surface is really needed.

Let us begin by considering some of the basic functional requirements of pulmonary gas exchange. At rest the oxygen needs of the body are relatively low, but in heavy exercise the muscles of an average human consume about 2.5 liters of oxygen every minute. In order to achieve this we ventilate the lung with about 30 breaths of over 2 liters per minute and our heart pumps some 16 liters of blood through the lung every minute. It turns out that under these conditions one red blood cell remains in the lung's gas exchanger for only about $^1/_2$ second, and in that time it has to pick up all the oxygen it can carry. This is why it seems likely that the lung is designed as a very efficient gas exchanger. Notice also that the lung must simultaneously eliminate a similar quantity of CO_2, since in heavy exercise muscle cells are predominantly fueled with carbohydrates (see Chapter 3), which results in a respiratory exchange ratio close to 1. Several arguments can be advanced that the uptake of O_2 by diffusion is the dominant function of the lung, and that the discharge of CO_2 is a secondary, though equally important function, so that in the following I shall concentrate on O_2 uptake.

What are the basic design features that allow the lung to perform its function as a gas exchanger efficiently? The most fundamental feature is that the lung is systematically organized in a hierarchical order of units that leads from macroscopic structures down to submicroscopic elements where gas molecules are exchanged between air and blood. This is first of all true for the pulmonary airways (Figure 4.1), which begin with a single tube, the trachea, and then branch in the form of an inverted tree, eventually reaching every point in space within the lung. The gas exchange units are something like the leaves on the outermost twigs of this airway tree which extend into the gas exchange tissue (Figure 4.2). Similarly, the pulmonary arteries that lead the blood into the lung begin as a single trunk at the right ventricle of the heart and then branch out into the lungs in the form of a tree that very closely matches and follows the airway tree. The pulmonary veins which collect the oxygen-enriched blood to bring it back to the left heart also form a similar tree, but its branches assume a position away from the

82 · · · Symmorphosis

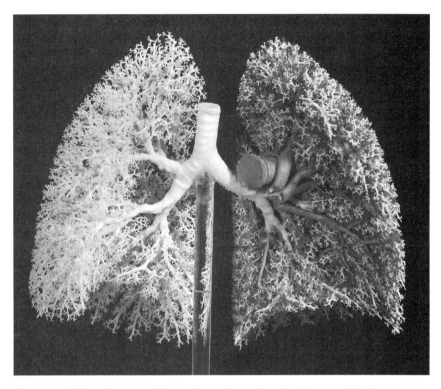

Figure 4.1 The pulmonary airways, here shown in a plastic cast of the human lung, form a branching bronchial tree with the trachea as its stem and the peripheral branches reaching into the gas-exchange parenchyma (see Figure 4.2). Pulmonary arteries and veins form similar trees whereby the arteries parallel the bronchial tree and the veins take a course between the broncho-arterial bundles. (From Weibel 1984.)

pulmonary arteries, about halfway between two airway units. These structures will be analyzed in more detail in Chapter 6.

The gas exchanger forms in the walls of the most peripheral branches of the airway tree (Figure 4.2), those branches that follow on the broken-off ends of the last twigs in the cast of Figure 4.1. Here the airway wall becomes scalloped by the formation of a large number of small side chambers, the alveoli, which number about 300 million in the human lung. Each of these alveoli is open to the central airway duct, but they are all densely packed to form a space-filling foam-like structure (Figure

Organ Design · · · 83

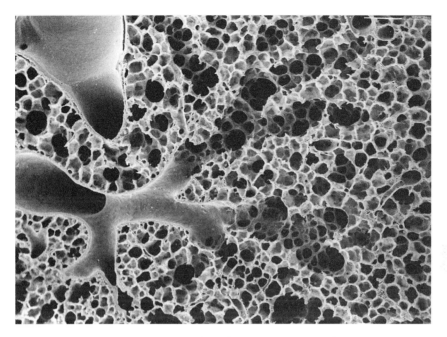

Figure 4.2 As the terminal airways, here shown in a scanning electron micrograph, reach the lung parenchyma, small side chambers, or alveoli, form in their wall.

4.3). Although the walls between alveoli are very thin, averaging less than 10 μm in thickness, they contain a dense network of capillaries that receive the blood from the end branches of the pulmonary arteries and drain it into the small pulmonary veins (Figure 4.4). As a result, the capillary blood occupies about half the volume of the interalveolar septa, as the walls between neighboring alveoli are called, so that only a very thin sheet of tissue remains to keep the blood separated from the air and to support this structure mechanically (Figure 4.5). This tissue barrier is about 100 times thinner than a book page, and yet it is itself composed of three layers of tissue (Figure 4.6): the alveolar epithelium, which lines the alveoli and is related to the airway epithelium, the capillary endothelium, which serves as the lining of peripheral blood vessels, and the interstitium, which contains connective tissue fibers as mechanical support.

Indeed, one of the most astonishing features of the lung is the fact

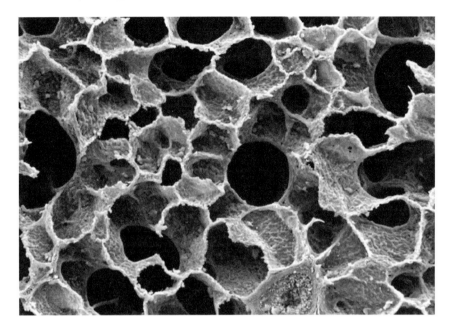

Figure 4.3 The lung parenchyma is made of densely packed alveoli which surround alveolar ducts as the end branches of the airway tree. It is a foam-like structure with thin walls. (From Weibel 1984.)

that only a very small amount of tissue, about 300 grams in weight, is used to build an organ about 5 liters in volume. This small amount of tissue must suffice to establish an orderly internal structure, to support blood in air over a very large surface, and to defend the lung against different adverse effects. This could well be taken as an expression of rigorously economic design, of sparing use of materials, but we shall see that the amount of tissue used is reduced primarily for other reasons.

This general description of lung structure has revealed the main features of the design of the gas exchanger: a very large surface of contact between blood in capillaries and air in alveoli combined with an extraordinarily thin tissue barrier. It is intuitively evident that these design features are favorable for gas exchange, but it is also clear that a number of serious problems may result from the fact that such a delicate layer of tissue is used to support the gas exchanger over the surface of a tennis court, problems that we will address in Chapter 5.

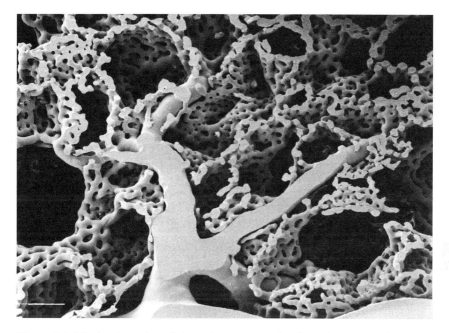

Figure 4.4 The last branches of the pulmonary arteries, here shown in a plastic cast, lead blood into a very dense capillary network which is contained in the walls between alveoli.

Modeling Gas Exchange in the Lung

In order to work out the relationship between lung design and gas exchange between air and blood, we must first grossly simplify the situation in a model that is limited to the essentials (Figure 4.7). In this model the branched airway system is reduced to a single tube through which air is moved in and out during ventilation. The 300 million alveoli are represented by a single chamber at the end of this tube. The dense network of capillaries becomes a single straight tube that lies close to the alveolar chamber from which it remains separated by a flat sheet of tissue, the air-blood barrier. Blood flows through this model capillary from the end branch of a pulmonary artery to the first collecting branch of the pulmonary veins. As the capillary blood flows along the alveolar surface—represented by the flat base of the alveolar cham-

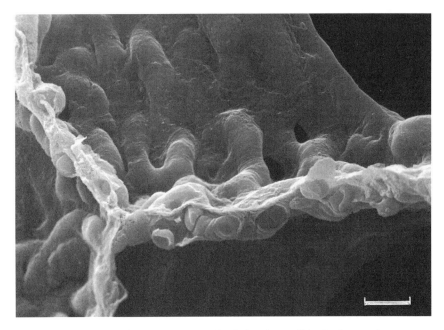

Figure 4.5 Scanning electron micrograph of alveolar wall in the human lung, showing erythrocytes in the capillaries and the very thin tissue barrier that separates air and blood. Scale marker: 10 μm. (From Weibel 1984.)

ber—it picks up O_2 from the alveolar air. This process occurs by diffusion through the tissue barrier and is driven by the difference between O_2 partial pressure in the alveolar air and in the capillary blood.

The graph at the bottom of Figure 4.7 shows the relationship between the partial pressures in air and blood. The O_2 partial pressure in alveolar air, symbolized as $P_{A_{O_2}}$, is on the order of 100 mmHg (or 13.3 kPa in SI units). This is the result of ventilating a closed chamber with fresh air which contains 21% O_2; on its way through the airways the inspired air becomes saturated with water vapor so that its O_2 partial pressure, $P_{I_{O_2}}$, is about 150 mmHg when it reaches the alveoli. The alveolar P_{O_2} is lower than this because only a small part of alveolar air is replaced by fresh air upon inspiration—we keep the largest part of this air in the lung when we exhale—and the extraction of O_2 by the capillary blood reduces the O_2 content of alveolar air. In addition, CO_2 is discharged from the capillary blood, which further dilutes alveolar oxygen.

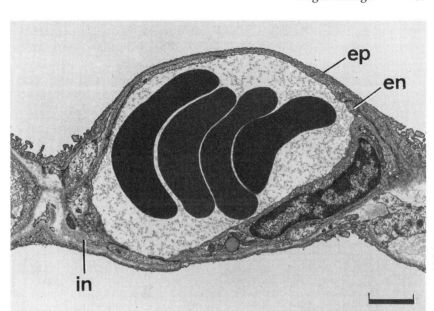

Figure 4.6 Alveolar capillary from the human lung shown in thin section to reveal the fine structure of the tissue barrier. This barrier is made of two thin cell layers, epithelium (ep) and endothelium (en), separated by a narrow interstitial space (in) with connective tissue fibers and cells. Scale marker: 2 μm. (From Weibel 1984.)

The blood that enters the capillary through the pulmonary artery is mixed *venous* blood which returns from the muscles and other parts of the body. Because these tissues have all extracted O_2 (Chapter 2), the O_2 partial pressure of mixed venous blood, $P\bar{v}_{O_2}$, is on the order of 20–40 mmHg, depending on the level of cell activity: closer to 20 in exercise, and about 40 at rest. Thus as the blood enters the gas exchanger, the driving force for O_2 diffusion from alveolar air into the capillary blood is on the order of 60 mmHg at rest and up to 80 mmHg in exercise. As O_2 is being taken up by the blood, the capillary O_2 partial pressure, Pc_{O_2}, gradually increases, and the driving force becomes smaller until it vanishes when the blood is equilibrated with alveolar air. The blood that leaves the lung through the pulmonary vein is *arterial* blood with an O_2 partial pressure Pa_{O_2} of about 100 mmHg. This blood is now transported to the muscles and the other organs in order to deliver O_2 for oxidative metabolism and pick up CO_2, only to return to the lung

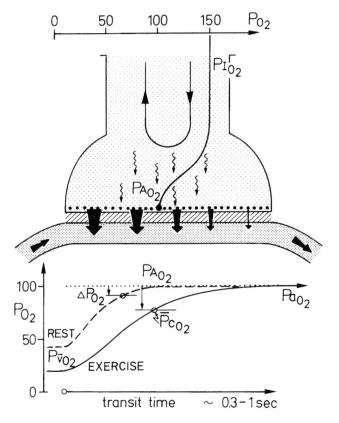

Figure 4.7 Model of pulmonary gas exchange. Oxygen tension in alveoli, $P_{A_{O_2}}$, is lower than in inspired air, $P_{I_{O_2}}$. As blood flows through the capillary, the P_{O_2} in capillary blood, $P_{c_{O_2}}$, gradually increases from mixed venous $P\bar{v}_{O_2}$ to arterial $P_{a_{O_2}}$ by equilibrating with $P_{A_{O_2}}$ across the barrier. With the higher rate of O_2 consumption and of blood flow in exercise, a greater proportion of the capillary path or transit time is needed for completing O_2 uptake than at rest. (From Weibel 1984.)

to discharge CO_2 and be recharged with O_2 in the next round of circulation.

The graph at the bottom of Figure 4.7 shows that the P_{O_2} of the blood increases along the capillary path in a nonlinear fashion. This is the result of the binding of O_2 to the hemoglobin contained in the red blood cells, a complex nonlinear process involving a chemical reaction between O_2 and hemoglobin. The graph also shows that, at rest, the blood is equilibrated with alveolar O_2 after less than half the capillary transit time of about 1 sec from arterial to venous end. In exercise, blood flow increases so that the transit time falls to about 0.3 sec, with the result that equilibration occurs only toward the end of the capillary path.

In view of modeling gas exchange in the lung, an important parameter is the mean capillary O_2 partial pressure, $\overline{Pc_{O_2}}$, obtained by integrating the curves in the graph. $\overline{Pc_{O_2}}$ determines the overall or average driving force for gas exchange in this lung model as the difference between alveolar and mean capillary P_{O_2}, and we see that this driving force is larger in exercise than at rest (Figure 4.7). This now lays the groundwork for deriving a simple formal model of O_2 uptake in the lung. Fick's law of diffusion states that the diffusive flow of a substance across a diffusion barrier is proportional to a driving force determined by the difference in partial pressure of O_2 between the two sides of the barrier, and inversely proportional to the diffusion resistance of the barrier. This is analogous to Ohm's law of electricity, where the current is proportional to the tension difference and inversely proportional to the resistance of the conductor. Applying Fick's law to gas exchange in the lung, the eminent Danish physiologist Christian Bohr derived, in 1909, the equation for O_2 flow rate

$$\dot{V}_{O_2} = (P_{A_{O_2}} - \overline{Pc_{O_2}}) \cdot D_{L_{O_2}}$$

where the term in parentheses is the overall driving force, as explained in reference to Figure 4.7. $D_{L_{O_2}}$ is the reciprocal of the diffusion resistance of the barrier—commonly called the diffusion conductance—for which Bohr introduced the term "diffusing capacity of the lung for O_2," which, as we will see, is most important for our considerations.

This equation tells us that the transfer of oxygen from alveolar air to the erythrocytes in the capillaries is a passive diffusive process driven simply by the partial pressures of oxygen in the two media separated by the barrier. Only under this condition can this equation be used. It is interesting to note that this has not always been an accepted condition. Earlier in this century physiologists maintained that the lung actively secreted oxygen into the blood. There were heated debates and it was only when Marie Krogh, August Krogh's wife, convincingly demonstrated in 1915 that the P_{O_2} in capillary blood was lower than that in alveolar air that the controversy subsided. Today the theory of passive diffusion is generally accepted for the lung.

A Large Surface and a Thin Barrier Determine the Gas Exchange Capacity of the Lung

With passive diffusion accepted as the mechanism for oxygen uptake, it becomes easy to define the nature of the conductance DL_{O_2}. The straightforward view, derived from basic physics, suggests that the conductance of the barrier, which is the reciprocal of the resistance to diffusion, is proportional to the surface area S, across which diffusion takes place, and inversely proportional to the barrier thickness τ, and that it also depends on the material properties of the barrier estimated by the permeation coefficient K:

$$D = K \cdot S/\tau$$

This conductance corresponds, in the electrical analog, to the reciprocal of Ohm's resistance, which is obtained as the ratio of length to cross-sectional area of the wire, multiplied by the resistivity of the conducting material.

Things are not so simple, however, for the reason that the pulmonary gas exchanger is made of several layers of tissue of different natures (Figure 4.6). Oxygen must first traverse the tissue barrier, which is made of two cell layers and an interstitial space, but for all practical purposes we can consider the entire tissue sheet to be a uniform diffusion barrier. The second layer to be traversed is blood plasma, and the third component is the population of erythrocytes contained in the capillaries. While tissue and plasma are simple barriers to be overcome by the diffusing molecules, the situation is different in the red blood cells where, in addition to movement by diffusion, O_2 molecules react chemically with hemoglobin. This is now a functional process entirely different from pure diffusion across the barrier. For this reason F. Roughton and R. Forster (1957) proposed subdividing the diffusing capacity into two serial components: a membrane diffusing capacity and an erythrocyte component. Since the membrane diffusing capacity,

DM, and that of erythrocytes, De, are in series we must add their reciprocals, that is, the respective resistances, to obtain the total diffusion resistance of the lung, or the reciprocal of the diffusing capacity:

$$1/D_L = 1/D_M + 1/D_e$$

For our purposes it is important to note that, even in this more sophisticated model, quantitative design properties are essential determinants of functional capacity. In Box 4.1 I have worked out the details of a model by which the morphometric properties of the lung are used to calculate a theoretical value of $D_{L_{O_2}}$. In short, the membrane diffusing capacity DM is a pure diffusion resistance determined by the permeability coefficient K for tissue and plasma, the gas exchange surface S(b), and the harmonic mean thickness of the diffusion barrier τ_{hb}, which corresponds to the diffusion distance from the alveolar surface to the red blood cells (Figure 4.7). Note that we are taking the barrier to comprise both the tissue and the plasma layers mainly because the flow of plasma, as fast as it is, is relatively slow compared with the diffusion velocity of oxygen. We then obtain the diffusing capacity of the membrane as

$$D_M = K_t \cdot S(b)/\tau_{hb}$$

which corresponds to the basic equation shown above, but applies only to the membrane barrier.

The second component, the erythrocyte diffusing capacity De, is not so easy to work out because the processes involved are quite complex and not well understood in every respect. The "solution" here is to choose a simple formula and to pack all the difficulties into a coefficient, which is given a Greek letter. Roughton and Forster suggested the following formula:

$$D_e = \theta_{O_2} \cdot V(c)$$

where V(c) is the capillary blood volume and θ_{O_2} the rate of reaction of whole blood with oxygen. In Box 4.1 several of the problems related to the estimation of θ_{O_2} are discussed. Most significant for us is that this

Box 4.1 Model for pulmonary diffusing capacity

The uptake rate of O_2 in the lung is described by the equation of Bohr (1909):

$$\dot{V}_{O_2} = (P_{A_{O_2}} - \overline{Pb}_{O_2}) \cdot D_{L_{O_2}} \tag{1}$$

where $P_{A_{O_2}}$ is the alveolar O_2 tension, $\overline{Pb}_{O_2}$ the mean capillary O_2 tension, and $D_{L_{O_2}}$ the O_2 conductance of the gas exchanger, or the pulmonary diffusing capacity for O_2. $D_{L_{O_2}}$ is determined, to an important degree, by design features, particularly by the extent of the gas exchange surface, the thickness of the air-blood tissue barrier, and the volume of capillary blood that can bind O_2 on its way past the air spaces.

The model that describes the effect of structural design on O_2 flow from air to blood, shown in Figure 4.7, considers the pulmonary diffusing capacity, $D_{L_{O_2}}$, to be the reciprocal of the overall resistance to O_2 diffusion, which can be broken into three serial resistances for the tissue barrier (t), the plasma layer (p), and the erythrocytes (e). The total resistance is the sum of the partial resistances, or the reciprocals of conductances, such that

$$1/D_L = 1/D_t + 1/D_p + 1/D_e \tag{2}$$

In this model the "membrane" diffusing capacity, D_M, is given by

$$1/D_M = 1/D_t + 1/D_p \tag{3}$$

Each of these partial conductances is determined by structural parameters that can be quantified morphometrically by stereological methods (see Box 4.2).

The first layer of the *membrane diffusing capacity*, the tissue barrier, can simply be considered as a sheet of variable thickness τ that separates two compartments, alveolar air and blood plasma, over an area S; thus Fick's law of diffusion determines O_2 flow across this barrier, and its diffusion conductance can be written as

$$D_t = K_t \cdot S/\tau \tag{4}$$

where K_t is Krogh's permeation coefficient of the tissue for oxygen and has a value of about 3.3×10^{-8} cm² min⁻¹ mmHg⁻¹.

As O_2 enters the barrier on its alveolar surface and leaves it through its capillary surface, it is reasonable to use the mean of these two surface areas, $(S(A) + S(c))/2$, as an estimate of the effective area of the tissue barrier. The effect of varying barrier thickness is that the conductance for O_2 will be variable from point to point—in fact, about inversely proportional to the local thickness. The effective overall barrier thickness is therefore the mean of the reciprocal local thicknesses or its harmonic mean thickness, τ_{ht}.

On the basis of these arguments, we define the conductance of the tissue barrier as

$$D_t = K_t \cdot (S(A) + S(c))/2\tau_{ht} \tag{5}$$

The second layer of the "membrane," the *plasma barrier*, can also be considered as a sheet of highly variable thickness separating erythrocytes from the endothelial surface so that its conductance can be estimated as

$$D_p = K_p \cdot S(c)/\tau_{hp} \tag{6}$$

with the permeation constant for plasma (K_p) being approximately the same as that for tissue.

In this model the membrane diffusing capacity is estimated as a two-step process. Several arguments can be advanced to consider the tissue and plasma layers as a single barrier separating air and erythrocytes; the main argument is that plasma is quasi-static because plasma flow is much slower than O_2 diffusion. In this case the diffusion-effective thickness of the barrier is the harmonic mean distance τ_{hb} from the alveolar surface to the erythrocyte membrane, so that D_M is estimated directly as

$$D_M = K_t \cdot S(b)/\tau_{hb} \tag{7}$$

A problem is how to estimate the "effective" total barrier surface $S(b)$. The most robust estimate is the alveolar surface area $S(A)$ or the mean of alveolar and capillary surfaces, but this may overestimate $S(b)$

if the hematocrit is low. In the normal lung this does not appear to be a problem.

The conductance of the erythrocytes, De, is of a different nature because it involves, besides O_2 diffusion, the rate of chemical reaction of O_2 with hemoglobin within the red cells. It is commonly expressed as the product of capillary blood volume, V(c), which can be measured morphometrically, and the rate of reaction of whole blood with O_2, θ_{O_2}:

$$De = \theta_{O_2} \cdot V(c) \tag{8}$$

The reaction rate θ_{O_2} poses a number of problems, as is reflected by the fact that estimates given in the literature vary greatly, with older estimates ranging from 2.7 to 0.9 ml O_2 ml^{-1} min^{-1} mmHg^{-1}.

It is generally accepted that the reaction rate θ_{O_2} is not a constant, but that it is affected by various blood properties, as expressed in the equation developed by Holland, van Hezewijk, and Zubzanda (1977):

$$\theta_{O_2} = k'_c \cdot (0.0587 \cdot \alpha_{O_2}) \cdot (1 - S_{O_2}) \cdot 0.01333 \cdot [Hb] \tag{9}$$

where k'_c is the initial reaction velocity of red cells when exposed to O_2 in solution (units: mmol^{-1} sec^{-1}); α_{O_2} is the Bunsen solubility coefficient of O_2; S_{O_2} is the initial (fractional) O_2 saturation of hemoglobin; and [Hb] is hemoglobin content given in g/100 ml.

The calculation of a reliable value for θ_{O_2} is fraught with a number of difficulties and uncertainties. On the basis of newer work, I have taken k'_c to be 440 mmol^{-1} sec^{-1} for all species.

A key problem of considerable importance is that k'_c is measured on desaturated blood, whereas the blood in the pulmonary capillary is 70%–98% saturated, and that k'_c falls rapidly above 75% saturation, so that θ_{O_2} is not a constant but falls as the blood travels from the pulmonary artery to the vein. The mean "effective" θ_{O_2} is about 60% of the value pertaining to desaturated blood. Estimated values of θ_{O_2} are given in Tables 4.1 and 4.2.

· · · ·

estimation depends on the hemoglobin concentration in the blood or, in morphological terms, on the capillary hematocrit, the volume density of erythrocytes in the blood.

The important conclusion from this model is that the functional pulmonary diffusing capacity is to a great extent determined by the design parameters of the lung's fine structure: the alveolar surface area S(A), the capillary surface S(c), the capillary volume V(c), and the harmonic mean thickness of the total barrier including tissue and plasma. To build a very large surface and an exceedingly thin barrier therefore makes for very favorable gas exchange conditions in the lung. To reduce the tissue mass to a small amount is thus not a question of economy, but rather one of optimizing design for gas exchange. In addition, the rate of O_2 binding θ_{O_2} is determined by the hemoglobin concentration or the hematocrit, the essential morphometric parameter of blood composition that plays an important role, as we have seen in Chapter 2. All these morphometric parameters can be estimated by stereological methods using an appropriate sampling strategy, as outlined in Box 2.2, if they are suitably modified to apply to the lung. In Box 4.2 the method for estimating the alveolar surface area is explained as a further example of this technique.

The Diffusing Capacity of the Human Lung

The results of such measurements on a set of healthy human lungs, obtained by Peter Gehr and Marianne Bachofen in 1978, are shown in Table 4.1. With a total lung volume of 4.3 liters at a medium degree of inflation, the alveolar surface area was found to measure 130 m²— which is about ¾ of a singles tennis court—with the capillary surface some 10% smaller. The capillary volume amounts to about 200 ml, which means that the contents of a wine glass are spread over the surface of a tennis court. Make a quick calculation dividing the capillary volume by the alveolar surface area and you will find that the layer of blood spread on the gas exchange surface is only about 1.5 μm thick, which is about half the thickness of a red blood cell! This is because the capillary diameter is about the size of a red cell and because oxygen can diffuse to the blood from both sides of the alveolar septum, as can be seen from Figures 4.5 and 4.6. Also, the capillaries form an extraordi-

Box 4.2 Stereological methods to estimate the morphometric parameters of diffusing capacity DL_{O_2}

The internal morphometric parameters of the lung are obtained by a sampling strategy, analogous to the methods outlined in Box 2.2, that begins with measuring the lung volume V_L. Tissue samples are then collected by a random procedure that gives all parts of the lung an equal chance of being chosen. These tissue blocks are processed for microscopy at different magnifications. The most important step in estimating alveolar surface area S(A) is to estimate the alveolar surface density at high magnification by electron microscopy.

To estimate this surface density, use a set of test lines of known length L and count the number of times I(A) this test line intersects the alveolar surface trace on the lung sections. The surface density is obtained as

$$S_V(A) = 2 \cdot I(A)/L$$

This equation is intuitively plausible if we consider the following thought experiment. If you stick a needle randomly into lung tissue, this needle will traverse some of the alveolar septa (see Figures 4.2 and 4.3). The number of intersections, I(A), will depend on the length of the needle, L, and on the size of the alveoli. The smaller the alveoli the larger the number of intersections obtained and the larger the surface per unit volume. The coefficient 2 is derived from theoretical considerations on the geometric probability of hitting a random surface in space with a needle.

The capillary volume density is estimated on the same sections by using a point-counting method (see Box 2.2). The total alveolar surface and the total capillary volume are obtained as the product of lung volume times alveolar surface density and capillary volume density, respectively.

The cascade sampling procedure (see Box 2.2 for more detail) refines this by taking the morphometric measurements in three steps, as illustrated in Figure B4.2 for the case of estimating the capillary volume and surface area. In electron micrographs magnified about 10,000× of alveolar septa (step 3), the septum is taken as the reference space. Using a stereological test system made of short lines of length l whose endpoints

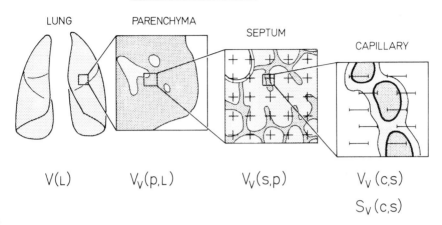

Figure B4.2 Model for morphometric analysis of lung by cascade sampling.

serve as test points, the intersections with the capillary surface I(c) and point hits on capillaries P(c) and septum (tissue + capillaries) P(s) are counted, yielding the capillary volume and surface densities (in alveolar septum) as

$$V_V(c,s) = P(c)/P(s)$$

$$S_V(c,s) = 2 \cdot I(c)/[P(s) \cdot (l/2)]$$

(Note that the total test-line length is the number of points in the septum multiplied by half the test-line length.)

In the second step, with the use of medium-power light micrographs, the volume density of alveolar septa in lung parenchyma is estimated, $V_V(s,p)$, and in the first step at low magnification the parenchymal volume fraction $V_V(p,L)$ is estimated, excluding large airways and blood vessels. The total capillary volume and surface are then obtained as:

$$V(c) = V(L) \cdot V_V(p,L) \cdot V_V(s,p) \cdot V_V(c,s)$$

$$S(c) = V(L) \cdot V_V(p,L) \cdot V_V(s,p) \cdot S_V(c,s)$$

The alveolar surface is obtained in the same way by counting alveolar surface intersections with the test line.

Finally, the *mean barrier thickness* is estimated using the electron micrographs of step 3. The micrograph is overlaid with a set of long test lines at random orientation. We measure the length of the barrier intercept by these test lines, that is, the distance from the alveolar surface to the erythrocyte surface for an estimation of total barrier thickness τ_{hb}. The harmonic mean barrier thickness is two-thirds of the harmonic mean of these intercept lengths.

· · · ·

narily dense network in the alveolar walls so that capillary and alveolar surfaces are nearly matched (Figures 4.4 and 4.5). This tells us that the gas exchanger is designed in such a way as to potentially allow the entire gas exchange surface to be involved in oxygen uptake: a very economic design.

Table 4.1 also contains estimates of different barrier thicknesses. For example, the tissue barrier has an average thickness of 2.2 μm; this arithmetic mean estimates the mass of tissue used to build the barrier: about 280 ml of tissue spread over 130 m². This average barrier thickness is, however, not relevant for gas exchange. The barrier is irregular in thickness, and over large parts the actual gas exchange barrier is quite thin, measuring less than 0.2 μm (Figure 4.6). Diffusion at each point of the surface is inversely proportional to local barrier thickness, so that O_2 will diffuse much more easily through the thinner parts of the barrier than through the thicker parts that contain connective tissue fibers. As a consequence, the relevant thickness for diffusion is the *harmonic* mean thickness of the tissue barrier (the mean of reciprocal local thicknesses), which is about 1/3 of the arithmetic mean, that is, about 0.6 μm. However, this is still incomplete. In the diffusing capacity model described above and in Box 4.1 we need to include the plasma layer in the estimate of functional barrier thickness; we then find the harmonic mean of total barrier thickness from the external membrane of the alveolar epithelial cell to the membrane of the erythrocytes to be 1.1 μm.

With all this information we can now calculate the conductances for the membrane, D_M, for erythrocytes, D_e, and for the total gas ex-

Table 4.1 Morphometric estimate of $D_{L_{O_2}}$ for young healthy adults of 70 kg body weight, measuring 175 cm in height

Morphometric data		(mean ±1 SE)		
Total lung volume (60% TLC)		4340	±285	ml
Alveolar surface area		130	±12	m²
Capillary surface area		115	±12	m²
Capillary volume		194	±30	ml
Air-blood tissue barrier thickness				
Arithmetic mean		2.2	±0.2	μm
Harmonic mean		0.62	±0.04	μm
Total barrier harmonic mean thickness		1.11	±0.1	μm
Conductances (ml/min/mmHg)				θ_{O_2}
Membrane	$D_{M_{O_2}}$	332		
Erythrocytes	$D_{e_{O_2}}$	383		
Total	$D_{L_{O_2}}$	178		

Source: From Gehr et al. (1978) and Weibel et al. (1993).
Note: TLC–total lung capacity; SE–standard error. "Effective" mean θ_{O_2}, taking into account its fall along capillaries, 1.8 ml/ml/min/mmHg.

changer, D_L, making suitable assumptions about the values for the diffusion coefficient K_t and the reaction rate of O_2 with blood θ_{O_2} (Table 4.1). It turns out that the membrane and the erythrocytes have about equal diffusing capacities; this means that these two serial components contribute equally to the diffusion resistance for oxygen. The theoretical value for the pulmonary diffusing capacity is on the order of 160–180 ml $O_2 \cdot$ min$^{-1} \cdot$ mmHg^{-1}. The lung thus has the capacity to transfer 160–180 ml of oxygen from the air into the blood every minute if the partial pressure difference for oxygen between alveolar air and capillary blood as the driving force for diffusion is 1 mmHg. In a human at rest oxygen uptake is about 300 ml/min so that a driving force of 2 mmHg would suffice (Figure 4.7). In order to transfer the 2.5 liters that a normal human being consumes in heavy exercise, a mean partial pressure difference of about 15 mmHg is required as a driving force. And in a well-trained athlete who can increase his oxygen consumption to some 4 liters or more, an even higher driving force would be needed, unless the athlete has adjusted his diffusing capacity to a higher value. As explained in relation to Figure 4.7, this increase in average P_{O_2} difference largely results from speeding up blood flow in exercise and thus shortening the time available for O_2 uptake into the erythrocytes.

How does this compare with physiological estimates of the diffusing capacity? In the physiological laboratory, particularly in relation to clinical physiology, the diffusing capacity is usually estimated in the resting individual. Typical values obtained are on the order of about 30 ml $O_2 \cdot$ min$^{-1} \cdot$ mmHg^{-1}, quite clearly much lower than what we predict as the theoretical or "true" capacity of the lung for oxygen uptake. If such estimates are made in exercising individuals, then considerably higher values on the order of 100 ml $O_2 \cdot$ min$^{-1} \cdot$ mmHg^{-1} are measured, but it appears that this is still lower than what we predict.

This difference between experimental and theoretical estimates of diffusing capacity has various possible explanations. Let us first note the meaning of the term "capacity": the maximum amount that can be contained or accommodated. Thus the capacity of a wine glass is not the amount of wine it actually contains but the quantity we can pour in until the first drop flows over the edge. As this example shows, capacity can only be fully exploited if the conditions are ideal; if the glass is even slightly tilted—or held in the hand—it cannot be filled to capacity. It is, therefore, not at all surprising that the "true" diffusing capacity of the lung cannot be fully exploited functionally. I shall discuss some of the possible limitations later on, for example in Chapter 5, where we will see that the alveolar surface is inherently unstable due to surface tension so that part of this surface may not be available for gas exchange all the time.

One way of estimating to what extent $D_{L_{O_2}}$ is used is to calculate the gradual loading of O_2 onto erythrocytes as they travel through the pulmonary capillaries from the arterial to the venous end, when they must have reached their arterial O_2 concentration (Figure 4.7); the procedure is called Bohr integration after Christian Bohr, who introduced these concepts. When this was done for human lungs, using physiological and morphometric data, my collaborators and I concluded that, even in heavy exercise, the arterial O_2 concentration is reached by the time the red blood cells have traveled through $2/3$ of the capillary length, so that the last third appeared redundant. This assumes, however, that all capillaries were perfused homogeneously. Of course, the data could

Organ Design · · · 101

also be interpreted to mean that only $^2/_3$ of all capillaries were fully perfused with the remainder underperfused, but that is an issue that cannot be discussed further for lack of physiological evidence.

At the same time we cannot exclude the possibility that the lung has a certain excess capacity or redundancy, perhaps as a safety factor. After all, the lung is in a very precarious situation, at the interface between the organism and the environment, the conditions of which we cannot usually control. For example, the P_{O_2} in ambient air depends on altitude, falling to rather low values in the higher elevations of the Rocky Mountains or in the Swiss Alps. It has been calculated that at high altitudes exercising athletes need all the diffusing capacity they have in their lungs to perform their heavy work. I shall pick this argument up again later in this chapter.

How Much Lung Diffusing Capacity Do We Really Need?

The estimates of pulmonary diffusing capacity of the human lung presented in Table 4.1 were obtained on seven healthy young adults, accident victims whose athletic prowess was not known. No physiological estimates, for example of $\dot{V}_{O_2}$max, were available on these individuals, but there was no indication that they had been athletes. We know, however, that well-trained athletes achieve a $\dot{V}_{O_2}$max that is 1.5–2 times higher than "normal" sedentary persons. Of course, it would seem that they could accommodate the higher O_2 uptake they require within the excess diffusing capacity that the normal healthy lung maintains. But it is still highly interesting to ask whether this is so, or whether athletes have larger lungs. This is of course not easy to study in humans, but a way out is to use comparative physiology, as we did for muscle in the previous chapters.

An early experiment of this kind involved the study of Japanese waltzing mice in comparison with white laboratory Swiss mice. Japanese waltzing mice have been raised as pets for centuries, mainly in China and Japan. They are exciting to watch because they incessantly twirl around their body axis at very high speeds, up to 2.5 turns/sec, as shown in the sequence of movie frames in Figure 4.8. In 1971 one of my students observed such mice in a pet shop and predicted that they must have a higher rate of oxygen consumption than the rather lazy Swiss

102 · · · Symmorphosis

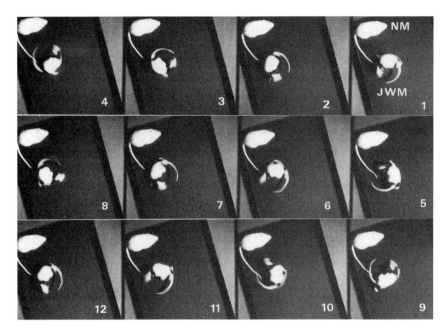

Figure 4.8 Twelve frames of a movie strip, representing ¾ sec, show the rapid "waltz" of a Japanese waltzing mouse (JWM) compared with a normal white mouse (NM). (From Geelhaar and Weibel 1971.)

laboratory mice and that the pulmonary diffusing capacity should therefore be proportionately larger. When she studied a group of these animals this is indeed what she found (Figure 4.9): average O_2 consumption per gram of body weight was 80% larger in waltzing mice than in white mice, whereas the pulmonary diffusing capacity estimated by morphometry was 60% higher, thus proving her predictions. This study was done over 25 years ago, and we were then not able to measure the limit of O_2 consumption $\dot{V}_{O_2}$max in these small animals. Indeed, I don't even know whether the waltzing mice would have been able to run straight on a treadmill. The best we could then do was to estimate average $\dot{V}_{O_2}$max over several hours of activity. From other studies we know that such average O_2 consumption is a little less than half $\dot{V}_{O_2}$max, so that the physiological results still have some meaning. They did allow us, at the time, to conclude that the higher rate of O_2

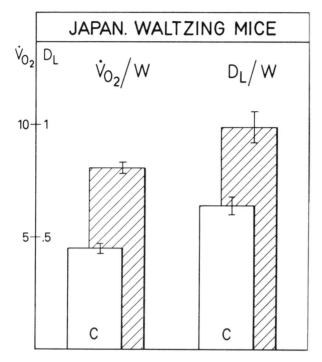

Figure 4.9 In Japanese waltzing mice both the rate of O_2 consumption and the pulmonary diffusing capacity are higher than in normal mice.

consumption in the waltzing mice was associated with a larger pulmonary diffusing capacity.

Later we performed, with a group headed by C. Richard Taylor at Harvard University, a well-controlled study comparing medium-sized dogs and goats. We have seen before that dogs have a $\dot{V}_{O_2}$max that is 2.5 times higher than that of goats. Did they have a proportionately larger diffusing capacity? As shown in Table 4.2 the structural differences between the dog and the goat lung were not very large: the gas exchange surface was only 15% larger in the dog and the capillary volume was about the same. The greatest difference was in the blood, since dogs have a particularly high hematocrit whereas goats have a low hematocrit and small red blood cells. As a first consequence the har-

Table 4.2 Morphometry of $D_{L_{O_2}}$ of pronghorn, dog, and goat

Parameter	Unit	Goat	Dog	Pronghorn
M_b	kg	27.70	28.20	28.40
$\dot{V}_{O_2}$max	ml · sec^{-1} · kg^{-1}	0.90	2.27	4.53
V_L/M_b	ml · kg^{-1}	63.80	56.30	163.80
$S(A)/M_b$	m^2 · kg^{-1}	2.43	2.75	7.37
$S(c)/M_b$	m^2 · kg^{-1}	2.13	2.17	6.50
$V(c)/M_b$	ml · kg^{-1}	4.38	4.18	10.87
τ_{hb}	μm	1.125	0.796	0.816
Hct	%	29.90	50.30	45.60
[Hb]	g/100ml	10.70	18.80	18.90
$D_{M_{O_2}}/M_b$	ml · min^{-1} · mmHg^{-1} · kg^{-1}	6.40	10.30	26.80
$D_{L_{O_2}}/M_b$	ml · min^{-1} · mmHg^{-1} · kg^{-1}	3.80	6.24	16.70
$D_{L_{O_2}}$ (28kg)	ml · min^{-1} · mmHg^{-1}	107.00	175.00	467.00

Source: Morphometric data from Weibel et al. (1987) and Lindstedt et al. (1991).

monic mean thickness of the diffusion barrier was 40% thicker in the goats because of a thicker plasma barrier resulting in a lower value of $D_{M_{O_2}}$ than in the dog. In addition, the difference in hematocrit resulted in a value of θ_{O_2}, 70% larger in the dog so that the dog's $D_{L_{O_2}}$ was 62% larger than that of the goat. Thus clearly the pulmonary diffusing capacity of the dog was larger but not by any means enough to match the difference in $\dot{V}_{O_2}$max. When we performed Bohr integration calculations, as explained above for the human lung (Figure 4.7), we found that the dog used all of its pulmonary diffusing capacity for oxygen uptake when exercising at $\dot{V}_{O_2}$max, whereas the goat did not (Figure 4.10). The goat had an excess diffusing capacity that was similar to the one we had estimated in humans, whereas the dog apparently made use of this excess capacity to achieve a higher rate of oxygen uptake. It was then interesting to find, in additional experiments, that the goat could easily run at $\dot{V}_{O_2}$max even when breathing air with reduced O_2 concentration, corresponding to hypoxia at high altitude, whereas the dogs did not tolerate any hypoxia. This is strong evidence that the observed excess diffusing capacity in the goat—and in humans—is real and can

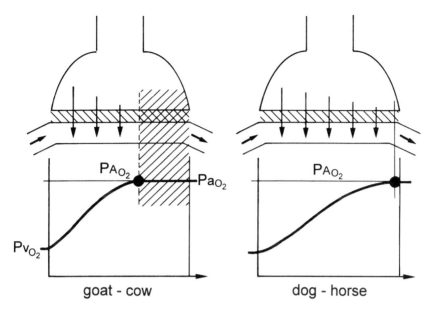

Figure 4.10 In the athletic animals, the dog and the horse, the entire capillary path length is needed in heavy exercise to complete O_2 uptake into the alveolar capillaries; in the sedentary animals, the goat and the cow, part of this path appears redundant. Compare with Figure 4.7.

be used under unfavorable environmental conditions to ensure adequate O_2 uptake.

The Gas Exchanger of the Athletic Pronghorn

What about the most athletic of all the mammals, the pronghorn (Figure 1.1), which is capable of running at a speed of 60 km/h for up to 40 minutes whereby it achieves a particularly high O_2 consumption rate: does its lung have an especially large pulmonary diffusing capacity? A number of years ago Stan Lindstedt, then at Laramie, Wyoming, studied the respiratory physiology of the pronghorn and found that its $\dot{V}_{O_2}$max achieved values that were five times those of a goat and two times those of a dog (Table 4.2). In Berne, we were able to study the lung of one of these pronghorns and to estimate its diffusing capacity.

The lung of the pronghorn is more than twice as large as the lung of a goat for animals of similar body size. When we estimated the pulmonary diffusing capacity we found that this lung had an internal surface area of 200 m², nearly three times that of a dog of the same size. The capillary volume was also much larger, whereas the barrier thickness was similar to that in a dog lung. The pulmonary diffusing capacity of the pronghorn was estimated at 470 ml O_2 · min^{-1} · mmHg^{-1}, thus about five times larger than in the goat and even three times larger than in the dog. Thus, clearly, this top athlete has a pulmonary gas exchanger that is commensurate to its needs for oxygen.

When we calculated by Bohr integration the fraction of the diffusing capacity that was used at $\dot{V}_{O_2}$max, we found that the pronghorn apparently maintains an excess diffusing capacity which is similar to that observed in humans and in goats. This may be astonishing because it is not easy to understand why an animal whose locomotor system is perfected to such an extent (Figure 1.1) and whose muscle mitochondria and capillaries are tightly adjusted to its energetic needs (Table 2.1) should be wasteful in building as large an organ as the lung. However, we may have been misled in this argumentation. This animal's normal habitat is in the high plains of the Rocky Mountains and it performs its phenomenal runs at altitudes between 2,000 and 3,000 m, where the ambient P_{O_2} falls as low as 100 mmHg compared with 160 at sea level. It is thus conceivable that the pronghorn needs a comparatively larger pulmonary diffusing capacity because it performs its heavy work under hypoxic conditions, where we can predict that the driving force for O_2 uptake by diffusion is smaller than at sea level.

This argumentation has recently found support when we studied another animal that notoriously lives and works hard under severely hypoxic conditions: *Spalax ehrenbergi*, the blind mole rat of Israel shown in Figure 1.4. When compared with white rats of similar body size, the mole rat achieved the same $\dot{V}_{O_2}$max when running on a treadmill under normoxic conditions in Massachusetts. And we found that the muscle mitochondrial volume was the same in both species, as we would expect. But the mole rat had a pulmonary diffusing capacity 1.4 times greater than the white rat; for the same $\dot{V}_{O_2}$max it was therefore excessive. However, we could then show, in studies done in C. Richard Taylor's laboratory, that the mole rat could achieve a higher "maximal" level of O_2 consumption than the white rat when they were running while breathing severely hypoxic air with an O_2 concentration as low as that prevailing in the burrows in Israel (11% O_2 instead of 21%). Thus we conclude that the larger diffusing capacity does not lead to higher $\dot{V}_{O_2}$max—which is determined by the amount of muscle mitochondria—but that it gives the animal a larger hypoxia tolerance, which is important for survival. It is conceivable that the pronghorn has a relatively larger lung for the same reasons.

The Effect of Reducing the Gas Exchanger

Another way of approaching the question of how much lung we really need is to surgically remove part of the lung and to see how well an animal performs with what is left. This intervention is frequently made in humans, for example when a tumor must be removed. With the intention of finding out whether the remaining lung can compensate for the lost part by growing larger, Connie Hsia and Robert Johnson at the University of Texas at Dallas performed partial pneumonectomies in adult dogs by removing the left lung, which makes up 45% of the total lung. One year later they found that the dogs achieved, with the remaining 55% of the lung, a $\dot{V}_{O_2}$max that was 85% of that before the operation. In a morphometric study they then found that the remaining right lung had expanded to fill the entire chest and that the gas exchange surface and the capillary volume were somewhat enlarged; however, these values were still smaller than those found in the combined right and left lungs of the normal dog. Indeed they found that the diffusing capacity $D_{L_{O_2}}$ reached 85% of the normal value, and thus the deficiency in $\dot{V}_{O_2}$max matched the deficiency in $D_{L_{O_2}}$. These investigators also measured physiological values of the diffusing capacity, using carbon monoxide as a tracer gas, and found that these values matched precisely the estimates of D_L obtained with the morphometric model (Figure 4.11). They concluded that the dog's pulmonary gas exchanger is completely used at $\dot{V}_{O_2}$max with no redundancy detectable; this, of course, agrees well with the findings reported above when we compared dogs with goats (Figure 4.10).

This experiment shows convincingly that the lung's gas exchanger is designed to supply O_2 to the organism at the rates required by heavy exercise, and that it may limit O_2 supply, at least in athletic species such

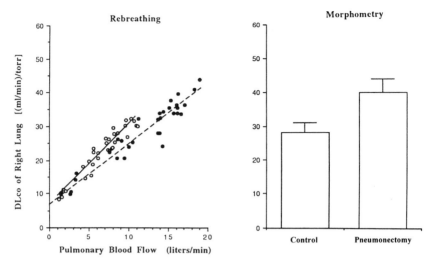

Figure 4.11 Estimation of the diffusing capacity of the right lung by a physiological method (left) and by morphometry (right) both show that after surgically removing the left lung (pneumonectomy), the diffusing capacity in the remaining lung increases by about a factor of 1.5 (solid circles) in comparison with unoperated controls (open circles), an augmentation that partially compensates for the loss of 40% of the total lung.

as the dog, specialized for high performance work. It also shows that the lung's capacity to adjust its structural design properties to altered functional needs is limited. A loss of 45% of lung structure in an adult animal is apparently not sufficient to cause the remaining lung to grow and reconstitute the original size of the gas exchanger, as was observed, for example, by Peter Burri when he performed pneumonectomies in growing rats, and as has recently been found by Connie Hsia in the dog when she performed lung resections in young puppies. The conclusions were different when, in later experiments by Hsia and Johnson, the right lungs, which constitute 55% of the total lung, were resected; in this case it was found that the residual smaller left lung not only expanded its volume but also grew new alveolar structures so that, after one year, the pulmonary diffusing capacity was restored to its normal size as found in the intact dog lung. These animals were also capable of achieving the same $\dot{V}_{O_2}$max as before the operation. I conclude from these experiments that the lung is capable of adjusting by compensatory growth the size of its gas exchanger to the body's O_2 needs, mainly during the period of growth. In the adult lung, however, a very large discrepancy between O_2 needs and gas exchange capacity is necessary to trigger true compensatory tissue growth.

Conclusion

We have seen that the lung's gas exchange function is to an important extent determined by structural design properties, particularly the combination of a large surface and a thin barrier. In athletic species such as the dog we have found that the pulmonary diffusing capacity, $D_{L_{O_2}}$, is closely matched to the $\dot{V}_{O_2}$max they can achieve, so that it can become a limiting factor for O_2 uptake and consumption under certain conditions. In other, more sedentary species, such as goats or humans, the lung appears to maintain a gas exchange capacity that somewhat exceeds that required at $\dot{V}_{O_2}$max. The apparent excess capacity is rather modest, however, amounting to no more than a factor of 1.5. Could this be a small safety factor built into the lung, perhaps because this organ lies at the interface to the environment? One argument in favor of this view is the finding that the most athletic North American mammal, the pronghorn, appears to maintain an excess capacity similar to that of goats if sea level conditions are considered; but the apparent excess vanishes under the hypoxic conditions that prevail at the high altitudes at which the animal lives and runs.

On the whole, the pulmonary gas exchanger indeed appears to be built on the principle of economy in that an exceedingly small amount of tissue is used to build and support a large gas exchange surface. The arguments for the development of such a thin barrier are not only those of economy, however, but also, or rather, those of efficiency of O_2 transfer from air to blood, which depends on making the diffusion barrier as thin as possible.

5: Problems with Lung Design: Keeping the Surface Large and the Barrier Thin

Building a human lung with an internal surface comparable to that of a tennis court and supporting it with a minimal tissue sheet that is 100 times thinner than a sheet of paper certainly provides advantages for gas exchange, as we have seen. But the resulting large diffusing capacity is obtained at a high price. We should note that the tissue has been minimized not for reasons of economy but, quite clearly, because a very thin tissue barrier is essential for the rapid diffusion of oxygen from the air into the blood: making the tissue barrier twice as thick would reduce the diffusing capacity by one quarter. The price for this thin barrier, however, is the acceptance of a number of potential dangers to the integrity of the system. These arise from fundamental problems for which special solutions must be found.

The difficulties derive from the fact that two different mechanical forces act on the very thin air-blood barrier. There is first of all a force from within the capillary, since the blood pressure that is required to drive the blood through the capillary network stresses the capillary wall and as a consequence tends to drive fluid out of the capillary. If this were not counteracted, fluid would gradually accumulate and cause the barrier to become thicker and thicker. Second, the barrier is under the action of surface forces from the air side; these develop at the alveolar surface and tend to collapse the alveoli.

The lung evidently has solutions for these problems, but they still remain as potential failures when conditions are unfavorable, as in certain diseases. Nonetheless we must ask whether the slight excess capacity we have observed in the lung's diffusing capacity is there in order to make sure that the oxygen supply will still be adequate even if some of the safety measures to counteract these problems should fail.

The stability of the alveolar surface is a serious problem because the surface of the air-blood barrier is fluid and forms an interface with air.

Surface forces are generated at this interface and will tend to flatten the surface. This is what happens when the waves generated by a drop falling onto water are rapidly smoothed. As can be seen in Figures 4.3 and 4.5, the alveolar surface is by no means flat. Curvatures occur at various levels, such as the hollow faces of alveoli or the bulging of capillaries on that surface. As explained below in more detail, the surface forces that are generated at this interface are proportional to the local curvature as well as to the prevailing surface tension. These forces will tend to flatten the capillaries and even to collapse the alveoli. They must be counteracted if collapse is to be prevented, but the question is how this can be achieved with so little tissue. The lung has found two solutions to this problem that complement each other: it provides a mechanical support through a clever design of the supporting fiber system, and it provides a specific surfactant for the control of surface tension.

A Fiber Continuum Supports Parenchymal Structures

The basic structural backbone of the lung, which keeps its shape and maintains the airways and blood vessels in proper position, is a system of connective tissue fibers strong enough to withstand a multitude of forces, for example, those that pull on all lung structures when we take a deep breath. It is typical of a fiber system that it must be under tension in order to serve as mechanical support: think of a spider web, or of a suspension bridge in which cables are extended between pylons to support the platform. Such a system of cables similarly suspends the alveolar walls with their capillaries in the lung air. To maintain the capillary bed expanded over a very large surface without disturbing gas exchange, however, requires a subtle, economic design of the fibrous support system. The solution of this problem appears ingenious: the lung is pervaded by a system of fibers that extends throughout the lung; this system is anchored at the hilum where the airways and blood vessels enter the lung from the central structures in the chest, and it reaches the visceral pleura, the outer membrane that enwraps the lung as a fibrous bag (Figure 5.1). This system forms a three-dimensional fibrous continuum that is structured by the airway system and is closely related to the blood vessels, for which it serves as direct support. All airways, forming

112 · · · Symmorphosis

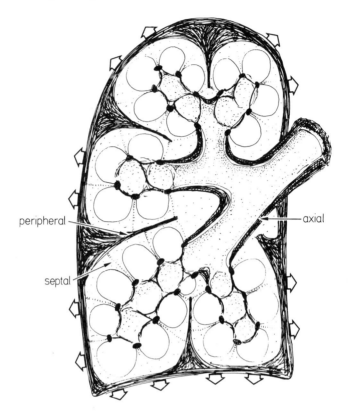

Figure 5.1 The fiber system of the lung is a continuum of connective tissue fibers made of three parts connected in series: the *axial* fibers that follow the wall of the airways out to the most peripheral branches in the lung parenchyma, the *peripheral* fibers that form the visceral pleura and extend into the lung parenchyma in the form of interlobular septa, and the *septal* fibers contained in the alveolar wall that extend from the axial to the peripheral fiber systems and thus support the capillaries. (From Weibel and Gil 1977.)

a tree (Figure 4.1), are enwrapped by a strong sheath of fibers that constitutes the *axial fiber system* of the lung; these fibers form the "bark" of the tree whose roots are at the hilum and whose branches and twigs penetrate deep into the lung parenchyma, following the course of the airways and eventually reaching every gas exchange unit. A second major fiber system is related to the visceral pleura; it is made of strong fiber bags that enwrap all lobes from where connective tissue septa penetrate into the lung parenchyma to separate, though incompletely, the hierarchical units of respiratory lung tissue: segments, lobules, and acini. We call these fibers the *peripheral fiber system* because they mark the boundaries between units. This peripheral fiber system also houses the major vessels, pulmonary veins, and arteries and binds them into the lung structure; it also contains the lymphatic vessels, to be discussed below.

The construction of the acinus—in many respects the functional unit of the lung parenchyma—is of particular importance for this discussion. The airway that leads into the acinus, the first-order respiratory bronchiole, continues branching within the acinus for 6–10 additional generations (Figure 4.2). These intra-acinar airways, called respiratory bronchioles and alveolar ducts, carry in their walls relatively strong fibers that are part of the axial fiber system and extend to the end of the duct system (Figure 5.2). But since the wall of intra-acinar air ducts is densely settled with alveoli, these fibers are reduced to a network whose meshes encircle the alveolar mouths (Figure 4.3). In engineering terms these fiber rings serve as anchoring cables for the network of finer *septal fibers* that spread within the alveolar walls; the other end of the septal fibers is attached to extensions of the peripheral fibers, which penetrate into the acinus from interlobular septa that form the boundary of the acinar unit (Figure 5.2). Thus a fiber continuum is constructed that extends along the airways through the septal fibers to the interlobular septa and the fibrous sheath of the pleura. It is hierarchically structured such that the fibers become systematically thinner, in both the axial and the peripheral fiber systems, as the gas exchange region is approached, so that the septal fibers are very thin and delicate. In this system there are no loose ends, and thus the fibers can be put under tension to serve their support function. This becomes particularly important in the alveolar walls, where a set of delicate fibers must carry the weight of the capillary blood which is suspended in air. This system is indeed comparable to the design of a suspension bridge on a microscopic scale, with the septal fibers as support cables and the axial and peripheral fiber systems as the two pylons.

The mechanical design of the alveolar septum deserves special attention because this is where gas exchange takes place (Figures 4.4 and 5.3). The relationship between fibers and capillaries therefore is par-

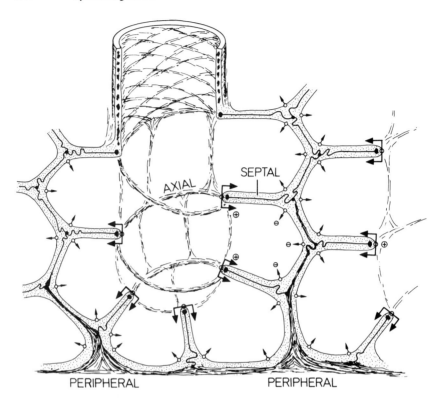

Figure 5.2 Simplified model of the fiber system in the pulmonary acinus to show the interaction between the three fiber systems in supporting the complex of alveoli in relation to the forces operating at the alveolar surface. Note that the surface forces (small arrows) are negative on the hollow surface of the alveolar chambers, but positive over the free edge of a septum (large double arrows), which would tend to flatten the surface and hence collapse alveoli. A major protective feature is the support of the positive tension on the free edge by the strong end fibers of the axial fiber system that surround the alveolar mouths. But this fiber support is not sufficient to stabilize alveoli. (From Weibel 1984.)

ticularly critical. The septal fibers form a single mesh that is interlaced with the—likewise single—capillary network as shown in the model of Figure 5.3b. As a consequence, the capillaries weave from one side of the septum to the other when the fibers are taut (Figures 4.4 and 5.3a). This arrangement has a threefold advantage: (1) it allows the capillaries to be supported directly on the fiber strands without the need for special braces; (2) it causes the capillaries to become spread out on the alveolar

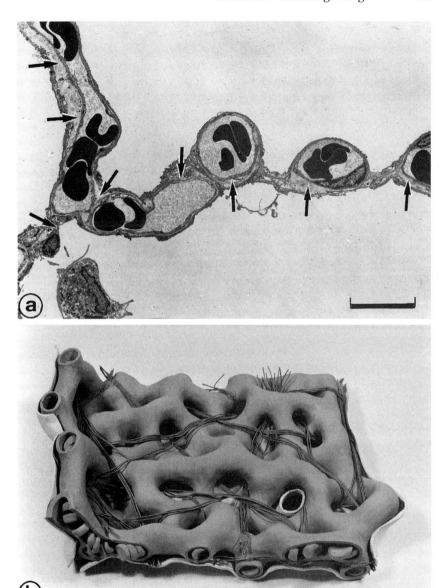

Figure 5.3 Design of the mechanical support of the alveolar septum with capillaries (compare Figure 4.5). The electron micrograph in (a) shows that the septal fibers, indicated by arrows, lie approximately in a plane. The bulging surface of the capillaries has a very thin barrier free of fibers. Scale marker: 10 μm. The reconstruction in (b) shows how the capillary network is interwoven with the lacework of the septal fibers. (b: From Weibel 1984.)

surface when the fibers are stretched; and (3) it optimizes the conditions for gas exchange by limiting the presence of fibers—which interfere with O_2 diffusion—to half the capillary surface. An interstitial space with fibers and fibroblasts exists only on one side of the capillary, whereas on the other side the two lining cells, endothelium and epithelium, become closely joined with only a single, common basement membrane interposed, as I shall discuss in more detail below.

In many respects, the fact that the pulmonary fiber system forms a continuum is an essential design feature of the lung. Its importance for the mechanical integrity of the gas exchanger becomes evident, for example, in pulmonary emphysema, which is characterized by a gradual destruction of tissue in the alveolar walls. This is often a consequence of the deposit of cigarette smoke particles in this delicate tissue; the deposits elicit an inflammation reaction that can lead to the enzymatic breakdown of tissue primarily if special defense mechanisms are lacking. When fibers are disrupted in this process they form loose ends and cannot be kept under tension; the broken fibers will retract, the septa break as a consequence, and larger air spaces form in the process of rearranging the fiber system in the damaged area. An irreversible loss of gas exchange surface is the inevitable result.

Controlled Surface Tension Determines Parenchymal Mechanics

Surface tension arises at any gas-liquid interface because the attraction forces acting between the molecules of the liquid are much stronger than those between the liquid and the gas. As a result the liquid surface will tend to be as small as possible. A curved surface, such as that of a bubble, generates pressure that is proportional to the surface curvature and to a surface tension coefficient. In a sphere the curvature is simply the reciprocal of the radius—so that the surface pressure is larger in a small bubble than in a large one, provided the surface tension coefficient is the same.

Now the lung is, in effect, a large assembly of small bubbles, the alveoli, that are all connected to the airways. In such a system the most critical effect of surface tension is that it endangers air space stability, because such a set of connected bubbles of different sizes is inherently unstable: in a closed system small bubbles should generate a larger surface pressure and consequently empty into the larger bubbles that then expand. The 300 million alveoli are all connected with one another through the airways and their size is variable (Figure 4.2), so why do the alveoli not all collapse and empty into one large bubble? There are two principal reasons.

1. The first reason is one of tissue structure. The alveoli are not simply soap bubbles in a froth; their walls contain an intricate fiber system, as we have seen. Thus when an alveolus tends to shrink, the fibers in the walls of adjoining alveoli are stretched, and this will, to a certain extent, prevent the alveolus from collapsing. It is said that alveoli are mechanically interdependent, and Jere Mead and others have argued that this interdependence can contribute to stabilizing the alveolar complex.

2. The second reason is related to the fact that the alveolar surface is provided with a special lining by a very specific surfactant, which has the peculiar property that its surface tension coefficient is variable: it falls as the surface is compressed and it rises when the surface expands. So when an alveolus tends to shrink because of surface tension, the surface tension coefficient becomes smaller and the surface force is reduced—whereupon the shrinking is halted. Conversely, when an alveolus expands, the surface becomes stretched and the surface force rises, because the surface tension coefficient increases. So shrinking alveoli have a relatively lower and expanding alveoli a relatively higher surface tension; as a result this tends to prevent the emptying of small alveoli into larger ones and contributes in important ways to the stability of the alveolar complex.

What is the nature of pulmonary surfactant that allows it to serve this vital function? The role of surface forces for lung mechanics was initially discovered in 1929 by Kurt von Neergaard, then professor of physical medicine in Zurich, but this did not attract much attention. The interest in the role of surface tension became greatly stimulated, however, in the 1950s following the discovery by Richard Pattle in England that froth collected from cut surfaces of the lung did not collapse. That froth coming out of the airways of some patients was very stable had been known in clinical medicine for a very long time and had even been used as a diagnostic criterion, but no one seemed to have drawn the important conclusion that the froth bubbles must be lined by

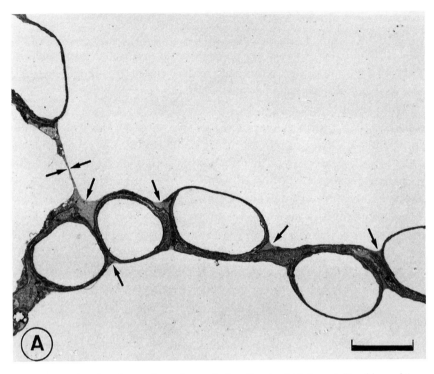

Figure 5.4 The alveolar surface of a perfusion-fixed rabbit lung is lined by a thin extracellular fluid layer which smoothes the surface by forming little pools in pits and crevices around the capillaries (arrows in A); it also spans across interalveolar pores (paired arrows in A). A higher-power view of a human lung specimen in (B) shows that the lining layer is a duplex structure with a fine (black) film of surfactant phospholipid overlaying the fluid pools of what is called the hypophase of the surfactant lining. (A: From Gil et al. 1979.)

a substance that has very low surface tension. Around 1960 John Clements then identified this substance as the di-saturated phospholipid dipalmitoyl lecithin, and he demonstrated convincingly that a film of this material spread on an aqueous surface, such as the alveolar lining, would provide the surface with dynamic surface tension properties. Through measurements of surface tension in vitro he demonstrated that the surface tension fell to near zero when the film area was reduced, and that it increased when it was expanded. This suggested that the surface tension coefficient would cycle between high and low values in the lung when alveoli expanded or contracted. Through an innovative micro-

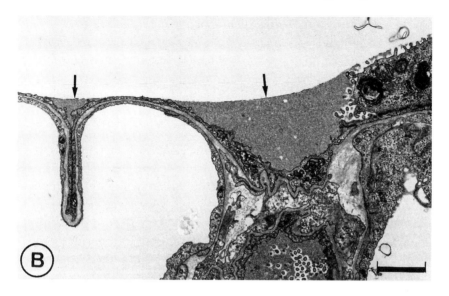

scopic method Samuel Schürch and John Clements later showed that this indeed also occurs in the alveolus.

Structures that support function are sometimes elusive. From 1962 on the evidence for the existence of a special surfactant lining of the lung was compelling. But its structural equivalent could not be found, so there remained a small grain of doubt about the true nature of this surfactant lining. In 1968, Joan Gil and I used a little trick, a fixation of the alveolar surface "from behind" by perfusing the lung vasculature with a suitable mix of fixatives, and this then revealed a duplex lining layer on top of the alveolar epithelial surface. It showed a layer of variable thickness we called the hypophase, which was overlaid by a very fine black film that had all the structural properties we would expect from a phospholipid monolayer (Fig. 5.4b). These studies also showed the reasons for the elusiveness of this layer: it is a very unstable film whose fluid hypophase is only loosely floating on the alveolar epithelium and is thus easily disturbed and washed off in standard preparation techniques for microscopy. Later it was found that the pulmonary surfactant lining is in fact a lipoprotein film, because a number of specific apoproteins associate with the phospholipid film and give it structural stability. This brought an interesting afterthought. Several years before the discovery of surfactant structure, I had found by electron microscopy in the lungs of humans and rats flakes of a strange material seemingly floating around the air spaces; it was made of regular arrays of dark and gray bands which I called tubular myelin. It later turned out that tubular myelin is one of the forms the lipoprotein structure of surfactant can assume and that the flakes I saw floating around had been surfactant material detached from the alveolar wall by the preparation procedure.

Figure 5.4 shows that the duplex lining layer of surfactant serves two distinct functions: (1) the aqueous hypophase tends to even out irregularities of the alveolar surface by accumulating mainly in the pits between capillaries, and (2) the film lowers surface tension very significantly because we see that the capillaries bulge toward the air space and are not completely flattened out by the surface force; also, we see in Figure 5.4a a fine film of fluid extending across an interalveolar pore, which is only possible if surface tension is very low.

Surfactant is a very specific substance, and we find that the alveolar epithelium contains a set of special secretory cells, known as type II cells (Figure 5.5), which are equipped with the organelles and enzyme systems to produce surfactant, both the phospholipid and the proteins. These relatively bulky cells are distributed sparingly over the alveolar surface, and they do not impair gas exchange significantly because they are tucked into corners where the barrier is thick anyway. They occupy no more than 5% of the surface. These type II cells take up their activity in the fetal lung at around the 28th week of gestation, so that a newborn human is provided with ample surfactant that allows the lung to be inflated and the alveolar surface unfolded on the first cry. In premature babies this may, however, not be so, and a severe respiratory distress syndrome develops if the lungs cannot be kept inflated. That this is due to a deficit in surfactant was the discovery of Mary Ellen Avery.

How important is surfactant for lung stability? How much stability could be achieved simply through the fiber continuum according to the theory of interdependence? To address this critical question Hans Bachofen performed experiments in excised perfused rabbit lungs. He inflated and deflated the lungs and measured the pressure needed to keep the lung expanded to a certain percentage of its total volume. The result of this was a pressure-volume curve as shown in Figure 5.6, and this confirmed the old observation that higher pressures are needed to inflate the lung from a collapsed state than to maintain inflation during

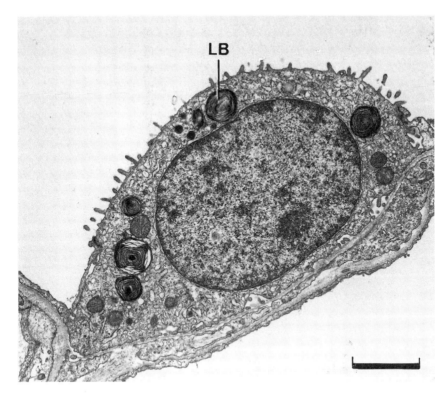

Figure 5.5 The alveolar epithelial type II cell from a human lung is a cuboidal cell that is well equipped with cellular organelles to serve as a secretory cell for surfactant phospholipids and apoproteins. The characteristic lamellar bodies (LB) are storage granules for the phospolipid, and the endoplasmic reticulum with ribosomes is the site of protein synthesis. Scale marker: 2 μm. (From Weibel 1980.)

deflation from maximal lung volume. We can interpret this curve immediately if we remember the essential property of surfactant: as the lung deflates, the alveolar surface is reduced, which compresses the surfactant film, and surface tension falls. Conversely, when the lung is inflated the surface and its surfactant film are expanded and surface tension becomes higher. The practical consequence of this is that we should keep the surfactant film well expanded, near the deflation curve of Figure 5.6, when breathing, because this allows us to maintain a large surface very cheaply. And indeed, that is what we do. We make use of this phenomenon when we occasionally take a deep breath, a sigh, so

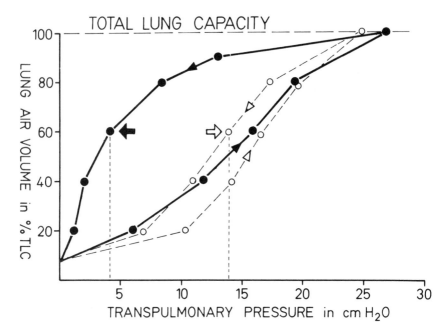

Figure 5.6 Pressure-volume curve of a normal air-filled excised rabbit lung (dark line) compared with that of a surfactant-depleted lung (fine broken line). Arrow heads mark inflation and deflation limbs. The surfactant-depleted lung collapses rapidly on deflation whereas the normal lung can maintain a large volume with little pressure on deflation. (From Bachofen, Gehr, and Weibel 1979.)

that our normal breathing is on the deflation slope of this pressure-volume curve and our respiratory muscles need to work less hard to inflate the lungs from breath to breath.

After this physiological experiment Bachofen fixed the lungs by vascular perfusion at 60% of total lung volume on the deflation slope and found that the alveoli were very nicely and regularly expanded as shown in Figure 5.7a. He then studied a second set of lungs in which he had first washed out the surfactant with a mild solution of detergent and then performed the same inflation-deflation maneuvers. Whereas the pressure-volume curve was similar on inflation, the lung volumes could not be maintained at low pressure on deflation: the lungs collapsed rapidly. To maintain 60% of total lung volume now required 14 cm water pressure compared with 4 in the normal lung (Figure 5.6). Fixation of these lungs revealed that the alveoli were collapsed into a

Problems with Lung Design · · · 123

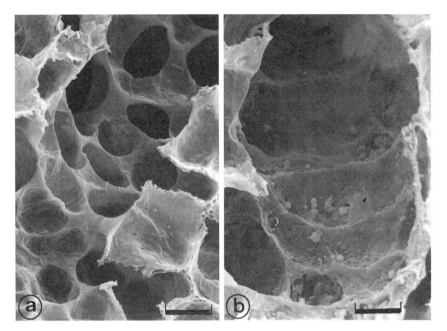

Figure 5.7 Scanning electron micrograph of normal air-filled lung (a) and surfactant-depleted lung (b) fixed at 60% on the deflation curve (arrows in Figure 5.6) shows open alveoli in (a) and collapsed air spaces in (b). Scale markers: 50 μm. (From Wilson and Bachofen 1982.)

flat patch between the axial fibers that appeared to be under great tension (Figure 5.7b).

We conclude from this experiment that surfactant is indeed essential for stabilizing alveoli. The tissue scaffold forms a platform that carries the capillaries and the alveolar epithelium, to which the surface lining layer with surfactant is loosely attached as a fine layer. By combining tissue tension with surface tension damped by surfactant, this structural system is molded into a stable gas exchange structure, where the capillaries, in particular, are evenly spread over the surface. It is quite evident that the concept of interdependence does not suffice at this fine structural level. However, if we consider the entire gas exchange surface of the lung, subdivided into 300 million alveoli, it is also evident that both a well-structured fiber system and a dynamic surfactant are needed to keep the large gas exchange surface expanded and in balance.

A key role here is played by the strong fiber rings around the alveolar mouths, the extension of the axial fiber system, that support the free edges of alveolar septa (Figures 5.2 and 4.3).

The balance of forces between tissue tension, surface forces, and capillary distending pressure also plays an important role in the fine tuning of the structure of the alveolar septum, in keeping the capillaries well expanded, and in maintaining good gas exchange conditions with a very thin delicate tissue barrier (Figure 5.4a).

Keeping the Barrier Dry and Thin

We have seen that the high diffusing capacity of the gas exchanger depends crucially on an exceedingly thin tissue barrier. This relationship is in jeopardy, however, because of the mechanical forces that act on the barrier. The chief problem is how fluid can be prevented from leaking excessively from the capillary and from either being soaked up in the tissue, so that the barrier becomes thicker, or even from flooding the alveoli. This danger is indeed real because the blood in the capillaries is under pressure from the fact that it is pumped in from pulmonary arteries where a mean pressure of about 20 mmHg prevails. This pressure is somewhat reduced when the capillaries are reached, but a positive pressure is needed to drive blood through the capillaries into the veins. This flow rate is very high during exercise, when over 10 liters of blood are pumped through the 200 ml of the capillary bed every minute. But there is an additional problem peculiar to the lung, the fact that capillaries also feel the weight of the blood column above them because the vascular system is a continuum suspended in air. This hydrostatic pressure affects mainly the capillaries in the lower parts of the lung which, as a consequence, are generally wider than those in the upper parts, and hence also better perfused.

All this results in a capillary distending pressure that stresses the capillary wall circumferentially. In contrast to solid tissues, the pulmonary capillaries are not encased in incompressible matter that can to some extent counteract this distending force. In the pulmonary capillaries only the delicate tissue barrier is available for this, and even the surface force is too low to contribute significantly. The stress on the capillary wall suggests that capillaries may burst, but this danger apparently does not exist under normal circumstances. The capillary wall appears strong enough to withstand these forces. However, it seems that it is not built with a very large mechanical safety factor. In thoroughbred racehorses the pulmonary arterial pressure can rise, during a race, to such values that capillaries rupture and cause bleeding into the lung.

Yet even under normal physiological conditions we might expect that fluid could leave the capillary through its wall, because blood plasma is under positive pressure. The barrier would become thicker as a result, and the alveoli would become flooded if the fluid oozed out into the alveolar surface lining layer. Clearly, this would significantly impair gas exchange. How is the design of the air-blood barrier adjusted to prevent this from happening?

The first and essential design property is that, as noted earlier, the barrier is made of three tissue layers: an interstitial space, related to the connective tissue fiber system, which is lined by two uninterrupted cell layers, one oriented toward the capillary and one oriented toward the alveolar space (Figure 5.8a). The capillary is lined by a complete lining of endothelial cells which are characterized by a very thin cytoplasmic extension that lines most of the capillary surface. The alveolar surface is lined by the alveolar epithelium, made of two cell types: type I cells, which are squamous cells with cytoplasmic extensions similar to those of endothelial cells (Figure 5.8a), and the cuboidal type II cells, which serve as secretors of surfactant (Figure 5.5). Although these two cell types occur in equal numbers, the type I cells cover 95% of the alveolar surface because the thin cytoplasmic leaflet of each cell covers an area of about 5000 μm (Figure 5.9). Thus both the endothelial and the epithelial linings are made essentially of broad thin cytoplasmic leaflets bounded by two plasma membranes and containing very few organelles besides cytoskeletal proteins (Figure 5.8b). These two leaflets constitute a minimal barrier to oxygen diffusion. However, their main structural elements, the four plasma membranes with their lipoprotein structure, are tight water barriers. In other words, oxygen can freely diffuse through these layers; water cannot.

A zone of danger exists, however, where the leaflets of two neighboring cells meet; here the membranes are attached to each other to form a seal at the so-called tight junctions (Figure 5.8c). But there is a difference between the two cell types. The epithelial seal is very tight, so that

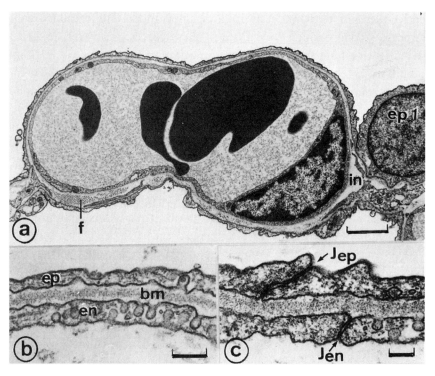

Figure 5.8 (a) Alveolar capillary with alveolar epithelial cell type I (ep1) with its broad cytoplasmic leaflets forming the outer lining of the tissue barrier and endothelial cell (en) lining the barrier on the inside. The interstitial space (in) is wide in the lower half of the capillary wall and contains fibers (f) and some cells, but thin with fused basement membranes (bm) in the upper part. This thin part is also shown at higher magnification in (b) and (c) with the intercellular junctions (J) shown in (c).

no fluid can be exchanged between the interstitial fluid space and the hypophase of the alveolar lining (Figure 5.4). In contrast, the endothelial junctions are leaky; they allow the transit of water and some solutes, including small macromolecules, from the plasma into the interstitial space. This is an important characteristic of capillary endothelia; remember that in Chapter 3 we examined the supply of glucose from the capillary by the small pore system in the endothelial seams. But this permeability also carries the potential danger of creating interstitial edema even under normal conditions.

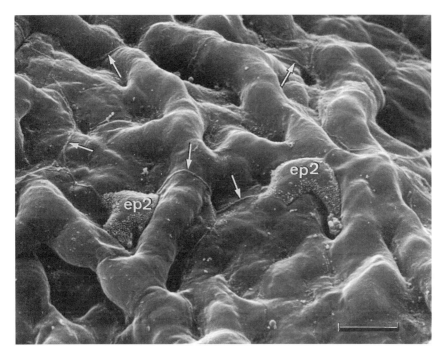

Figure 5.9 The surface of the alveolar wall in the human lung, seen by scanning electron microscopy, reveals a mosaic of alveolar epithelium made of type I and type II (ep2) cells. Arrows indicate the boundary of the cytoplasmic leaflet of a type I cell which extends over many capillaries. The boundary is made by intercellular junction lines (Figure 5.8c). (From Weibel 1980.)

How is this potential danger avoided? Two points are noteworthy. First, the design of the interstitial space is a clever one. Each of the two cell layers is provided with a basement membrane that is tightly bound to the cell membrane by special adhesion molecules. The basement membranes provide the cell lining with some mechanical strength in addition to that offered by the cytoskeleton. Now it is found that the epithelial and endothelial basement membranes are fused over half of the barrier surface (Figure 5.8b), and this occurs in the regions of the capillary that are not associated with fibers (Figures 5.3 and 5.8a, b). As a result, no interstitial fluid can accumulate in half of the barrier. This portion of the barrier is therefore kept dry and thin purely by design. In other words, half of the gas exchange barrier, and that part which is

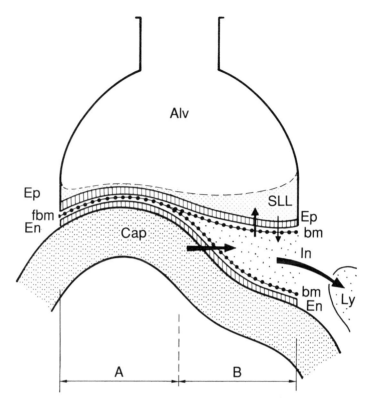

Figure 5.10 Model of the alveolar-capillary tissue barrier in terms of its fluid spaces. Over half the barrier the endothelial (En) and epithelial (Ep) basement membranes (dotted lines) are fused (fbm), thus excluding interstitial fluid. In the other half the interstitium (In) is associated with the septal fiber system and is open toward the connective tissue around the peripheral fiber system where lymphatics (Ly) can drain off the excess fluid, keeping the barrier dry and thin. The surface lining layer (SLL) is unevenly distributed over the surface. The size of the arrows indicates the relative intensity of fluid flow.

thinnest to begin with and thus the most effective for gas exchange, is protected from losing its main features as a gas exchanger by the elimination of the interstitial space from that part of the barrier. This is shown schematically in Figure 5.10. Thus, fluid can accumulate only in those parts of the barrier associated with fibers. This is advantageous:

these are the thicker parts of the barrier anyway, so that a further thickening has a comparatively small effect on gas exchange.

But now a second point is important to note: this interstitial space is an open system, bounded by the two layers of endothelium and epithelium like a sandwich, but open "sideways" along the fibers so that the fluid it contains can flow out of the gas exchange region. The fibrous continuum outlined above (Figure 5.1) extends from the alveolar septum to the peripheral fiber system (Figure 5.2) and thus provides an escape route or drainage pathway for interstitial fluid toward the larger connective tissue masses. These surround blood vessels and form interlobular septa where the lymphatics are found that drain the fluid out of the lung and eventually back into the blood (Figure 5.10).

Thus the impressive features of this design for the protection of the gas exchange function are that the thin barrier portion is made inaccessible to interstitial edema by fusion of the basement membranes, and that the thick parts are provided with an efficient drainage pathway along the fiber system. The other important feature is the different permeability properties of the two cell layers of epithelium and endothelium, such that the leaky endothelium allows the interstitium to be flushed whereas the tight epithelium protects the delicate surface lining layer from the danger of any uncontrolled inflow of fluid from the capillaries and interstitium.

Even though I have stressed the extraordinarily clever design features of the septum by which the lung defends itself against untoward effects of alveolar collapse or flooding of the barrier spaces, we should remember that the absolute prerequisite for all this to really function is the intactness and the vitality of the cell layers lining the barrier. Without live and active type II epithelial cells, for example, there is no surfactant production and the lung goes into respiratory failure, as happens in the prematurely born baby who develops what is called the respiratory distress syndrome of the newborn due to inadequate surfactant production. Clinicians are now trying to overcome this quite frequent condition by supplying exogenous surfactant through the airways.

The delicate type I epithelial cell is extremely vulnerable, and there are multiple conditions in which it can be destroyed. Shock, affecting, for example, accident victims with multiple bone fractures, is perhaps the most common. These patients can on admission to the emergency room rapidly develop adult respiratory distress syndrome, caused by a profuse flooding of large parts of the lung with fluid from damaged alveolar capillaries. The most severe damage is that of the type I epithelial cell, because it is not easily repaired, whereas endothelial cells rapidly form new linings when they are damaged. Such pathological developments help demonstrate how important the integrity and vitality of the cells constituting the lung's delicate tissue is for maintaining adequate lung function.

Conclusion

We have seen that structural design plays a major role in ensuring stable lung function in spite of severe mechanical problems stemming from the fact that the gas exchanger must be suspended in air over a very large surface with scant tissue support. The cell linings of the barrier are specifically designed to keep the barrier dry and thin. A particularly apt bioengineering design of the fibrous support structures provides a scaffold strong enough to ensure structural integrity and thin enough, in the critical region, to be no impediment to gas diffusion. But only the addition of a new and highly specific design element, the surfactant lining of the alveolar surface, allows the tissue elements to be minimized and still be adequate to stabilize the very large and complex inner lung surface. As we have seen, to minimize tissue in this case is not primarily a matter of economy, but rather one of providing a form of the gas exchanger that allows optimal conditions for the performance of the gas exchange function.

6: Airways and Blood Vessels: Ventilating and Perfusing a Large Surface

In Chapter 4, I argued that the excess capacity of the human pulmonary diffusing capacity could be conceived as a safety factor. This seems possible because our model supposed ideal boundary conditions for gas exchange: for the diffusing capacity to be fully used in gas exchange one of the central—though not explicitly stated—assumptions was that the large surface was homogeneously perfused with blood in all parts of the dense capillary network all the time; and that all points on the alveolar surface have at all times the same O_2 partial pressure as the driving force. But it is certainly unlikely that such ideal conditions can be achieved in an organ as complex as the lung. Any deviation from this ideal state, any mismatch between ventilation in the alveoli and capillary perfusion, will reduce the functional use that can be made of the theoretical diffusing capacity, which would explain why the physiological $D_{L_{O_2}}$ is smaller than the morphometric or theoretical one. However, remember that the dog and the horse seem to be able to use all of their $D_{L_{O_2}}$ when exercising at high levels, which suggests that the lung has good solutions for the apparent problems in ventilating and perfusing the large gas exchange surface. What are the problems and how have they been solved?

One of the foremost problems is how to ventilate with air and perfuse with blood a surface about the size of a tennis court—and to do that efficiently, homogeneously, and with perfusion and ventilation well matched. This is a demanding task of engineering. The key problem is that the bolus of fresh air that we inhale upon ventilation enters the lung at the bifurcation of the trachea in the center of the chest; likewise the blood that is to perfuse the capillaries is ejected as a bolus from the right ventricle into the pulmonary arteries, and this occurs also in the center of the chest. From there, air and blood must be distributed to all points on the alveolar surface as evenly as possible. Let us imagine the problem. If you stand in the middle of the tennis court and wish to evenly sprinkle the surface with water, the distance that your water-jet must cover varies from a few cm to up to 10 m—an impossible job.

An obvious solution to this problem is to fold the surface up in order to bring it closer to the origin of the airways and vessels; in a way, what needs to be done is to crumple a sheet of 130 m² into a tight ball of about 4 liters volume. But simple "random crumpling" will not do. It must be done in an orderly fashion so as to ensure each point on the surface direct access to inflowing air. This occurs during morphogenesis of the lung through the systematic development of a branching tree of airway tubes that expands as the lung grows and eventually forms the gas exchanger units at the distal twigs of the tree. Looking at this important process in some detail will allow us to understand how form is developed to serve function best.

Morphogenesis of Airways, Vessels, and Gas Exchanger

Lung morphogenesis follows a number of principles that are all favorable to the construction of a well-functioning gas exchanger:

- lung morphogenesis proceeds from the center to the periphery, adding unit after unit as the lung grows;
- it uses simple epithelial tube structures to build a branched system of channels, both in the airways and in the blood vessels, that connect the distal parts to the center in a natural hierarchy;
- formation of the airway tubes sets the pace for lung development;
- formation of blood vessels occurs in tight association with the airway system, which results in three matched interdigitating trees that converge in the distal gas exchange region;
- only the most distal airways and vessels are transformed into gas exchange structures, keeping all the central ones as pure distribution pipes.

The key morphogenetic process is the stepwise growth of the airway tree by dichotomous branching: the terminal twigs of the tree grow in length, and then each of their tips splits into two buds, giving rise to two new branches (Figure 6.1). With every generation of branching the number of tubes is thus doubled. This process has one important char-

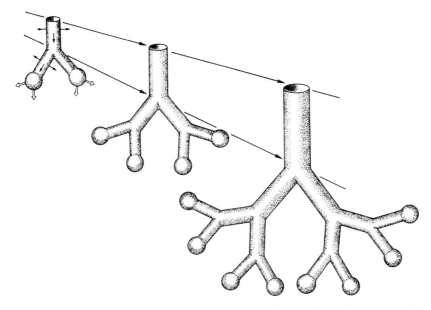

Figure 6.1 Model of airway growth, showing branching at the tip (open arrows) and the growth of the branches (small arrows).

acteristic for our discussion. The growth of the airway surface occurs not by simple expansion, like the inflation of a balloon, but rather as a process of progressive internal subdivisions of the growing lung. As a result a large surface becomes packed into a limited volume and the distance from the origin to the surface is kept short all along this growth process, which ends only when body growth subsides.

Toward the end of lung development in the fetus, and to an important degree even after birth, the wall of the most peripheral airway generations becomes further folded up by the formation of alveoli, a process that approximately doubles the surface area without changing the volume (Figure 6.2). The 300 million alveoli are arranged around the peripheral airway ducts, with the result that their gas exchange surface is very close to the ventilating airway system (Figure 4.2). Finally, the alveolar surface is further increased by the impressions of the capillaries (Figure 4.5), the last step in folding up the gas exchange surface.

Thus the large gas exchange surface is, through its design, brought

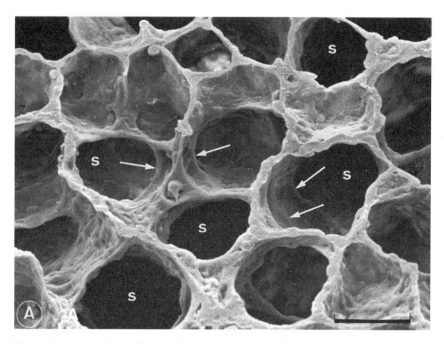

Figure 6.2 Formation of gas-exchange structures in the rat lung. (A) At day 6 after birth the peripheral airways are still in the form of saccules (s), the widened tubes of the last few generations which have thin walls to serve as gas-exchanging units. (B) On day 17 these saccules have become subdivided by enlargement of the secondary septa seen as low ridges in (A), where they are marked by arrows; small alveoli (s) have now formed as side chambers of the saccules. (From Burri 1997.)

closer to the source of fresh air so that the distance for ventilation from the center of the chest to every point of the alveolar surface is on the order of only 15–30 cm, or 40–55 cm from the mouth and nostrils. This is nearly two orders of magnitude shorter than the sprinkler distances we estimated for the tennis court.

The next problem is how to gain access to this surface with a system of blood vessels that allows even perfusion of the surface from a central point in the chest. The solution is again found during morphogenesis. Very early in lung development a network of blood vessels surrounds the primitive epithelial airway tube. As the airway tube develops and branches, this vascular network remains associated with its surface and

Airways and Blood Vessels · · · 135

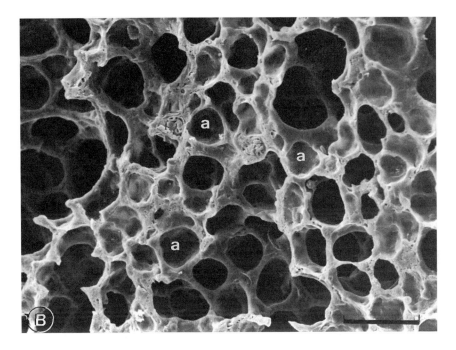

therefore grows strictly in parallel with the surface of the tube system. Arterial branches develop from this network, and these closely follow the course of the branching airway tree so that they eventually form an arterial tree that is strictly parallel to the airways (Figure 6.3) and develops additional branches only in the very periphery where the airway tube forms alveoli. Because of this arrangement the gas exchange units of the lung are broncho-arterial units, with the pulmonary veins collecting the blood between these units. This design provides the engineering basis for matching ventilation and perfusion in all units of the gas exchange surface.

Designing the Airway Tree for Efficient Ventilation

The result of morphogenesis is an airway tree whose branching pattern strictly follows dichotomy. With each generation z the number of branches is doubled so that the number of branches follows a power

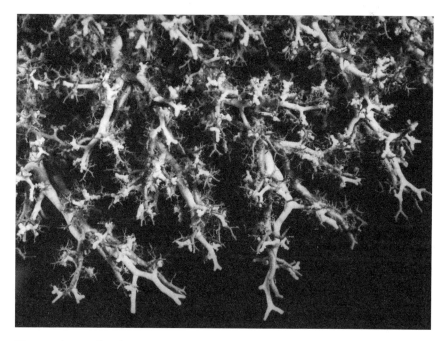

Figure 6.3 Peripheral airways and the accompanying branches of the pulmonary artery in a resin cast of the human lung.

law: $N(z) = 2^z$. In the human lung we find that the airways branch over 23 generations on average, which results in about eight million terminal branches (Figure 6.4).

A further result of morphogenesis is that the ends of the tree must reach into every corner of the lung, and this means that the branching pattern must, in its details, be adjusted to the available space. Because the shape of the lung is determined by the chest wall and the diaphragm, as well as by the heart, this introduces irregularities in the branching pattern; some branching paths will have to stop before 23 generations, while others continue to branch beyond that average value. The range of generations is between 18 and 30 in the human lung.

In order to study the details of the branching pattern, let us look at resin casts of human airway trees such as the one in Figure 4.1. The detail in Figure 6.5 shows that the dimensions of the airway branches are reduced with each generation in length as well as in diameter. This size reduction is again irregular in that the two sibling branches of one

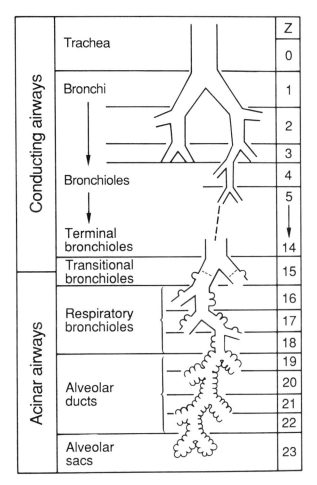

Figure 6.4 Organization of the airways by functional zones in relation to generations Z of dichotomous branching. (After Weibel 1963.)

parent may be of different size. This irregularity, however, is superimposed on a basic pattern which follows the principle that daughter branches are always smaller than the parent branch, and this is maintained throughout the airway tree, in spite of irregularities. There is, in the whole tree, not one daughter branch that is larger than its parent.

To understand the engineering principles underlying the construction of the airway tree, it is best to uncover the basic pattern first. Let us then

Figure 6.5 Peripheral airways from human lung in resin cast. Note similar branching pattern at all levels.

look for a model that predicts how a branching airway tree must be constructed to ensure the most efficient ventilation. Such a model is represented by Murray's law, which was worked out for blood vessels but which can in principle also serve for airways. We encountered the historical background of this law in Chapter 1, where the principle of economy was discussed with reference to the study of W. R. Hess in 1913, in which he concluded that the dimensions of blood vessels tend to minimize the work of transport (Figure 1.5). Hess predicted that the diameter of blood vessels is reduced at each branch point by a factor of the cube root of ½. When he measured the succession of branch diameters in larger vessels of the neck and in the retina (Hess practiced as an ophthalmologist in a small town before becoming professor of physiology at the University of Zurich), the results came very close to this predicted value.

The pioneering work of Hess was later refined by C. D. Murray (1926) with a slightly different analysis. What is now called Murray's law says that maximal efficiency of blood flow in a vessel is achieved if the blood flow shows a constant relation to the cube of the vessel radius—which basically confirms the main result of Hess. The arguments that led to this general conclusion are outlined in detail in Box 6.1. Although some of Murray's arguments are contested, his general conclusions are still valid; they can also be arrived at by different but related arguments. Murray worked out the minimum cost of work for overcoming flow resistance and maintaining a certain blood and vessel volume. This fundamental law of maximal efficiency leads to an engineering rule that determines the size of the branches in a vascular tree in which the flow through the two daughter branches must be the same as the flow through the parent branch. As a result of optimization considerations it is found that the radius (or the diameter) of the daughter branches must, on average, be reduced by a factor of $\sqrt[3]{1/2} = 2^{-1/3} \sim 0.79$ with each generation of branching. (This factor indeed seems to be an accepted rule of engineering, because it also determines the scale by which pipes are dimensioned when constructing the gas pipeline system for a city.) In recent years, it has been shown by more detailed analysis that this factor is a first approximation and that the true factor should be not $2^{-1/3}$ but rather $2^{-1/2.7} \sim 0.77$, but that does not change the principle.

Murray worked out these relations for blood vessels, so the question arises whether this law can also be applied to airways. Clearly, the flow resistance in conducting airways is also governed by Poiseuille's law and will therefore fall with the fourth power of the radius. But the equivalent to the cost of blood is not so evident, largely because we do

Box 6.1 Murray's law of the economic design of blood vessels

In 1926 C. D. Murray published the first in a series of papers in which he worked out the design conditions that would ensure the economic design of blood vessels, based on what he called the "physiological principle of minimal work," which he postulated to be one of the fundamental principles of physiological organization; this principle had already been used by W. R. Hess in 1913 in his analysis of the same problem. Murray argued that the size (radius) of blood vessels should be adjusted to minimize (1) the cost of work caused by the resistance to the blood flow and (2) the costs of making and maintaining the mass of blood in the vasculature (he also included in the latter the cost of building and maintaining blood vessel walls, which he took to be similar to that of the blood).

The cost of blood transport is the product of blood flow $\dot{Q}$ and the pressure p needed to overcome the vascular resistance, which is determined by Poiseuille's law

$$p = \frac{\dot{Q} \cdot l \cdot 8\eta}{\pi \cdot r^4} \tag{1}$$

where l is the length, r the radius of the vessel segment, and η the blood viscosity.

The cost of blood (and of vessel wall) he took to be given, as the vessel volume v multiplied with a "cost factor" b

$$b \cdot v = b \cdot l \cdot \pi r^2 \tag{2}$$

so that the total work to be done is

$$E = p \cdot \dot{Q} + b \cdot v = \frac{\dot{Q}^2 \cdot l \cdot 8\eta}{\pi \cdot r^4} + b \cdot l \cdot \pi r^2 \tag{3}$$

In equation (3) the first term falls with the fourth power of the vessel radius r, whereas the second term rises with the square of r. The combined function has a minimum where the derivative dE/dr is zero (see Figure 1.4). Differentiation and setting to zero yield

$$\dot{Q}^2 = \frac{\pi^2 \cdot r^6 \cdot b}{16 \cdot \eta} \quad \text{or}$$

$$\dot{Q} = r^3 \cdot \sqrt{\frac{\pi^2 \cdot b}{16 \cdot \eta}} = k \cdot r^3$$

since all the terms in the root are constants.

Murray (1926a) concluded from this that "one of the simplest requirements for maximum efficiency in the section [of the vessel is] that the flow of blood past any section shall everywhere bear the same relation to the cube radius of that vessel at that point." This is what is now called "Murray's law."

In a subsequent paper of the same year (1926b) Murray extended this law to systems of branched vessels as they characterize the vasculature of the lung and of the systemic circulation. The principle is simple: the flow through the two daughter branches of radius r_1 and r_2 must be the same as through the parent branch of radius r_0:

$$\dot{Q}_0 = \dot{Q}_1 + \dot{Q}_2 \quad \text{so that}$$

$$k \cdot r_0^3 = k \cdot r_1^3 + k \cdot r_2^3 = k \cdot (r_1^3 + r_2^3) \quad \text{or}$$

$$r_0^3 = r_1^3 + r_2^3$$

In symmetric trees where $r_1 = r_2$, this reduces to

$$r_0^3 = 2 \cdot r_1^3 \quad \text{or}$$

$$r_1 = r_0 \cdot \sqrt[3]{1/2}$$

from which we see that the daughter branches are smaller than the parent branch by a factor of the cube root of ½. This is also called Murray's law.

The same result had, in fact, been derived independently in 1913 by Walter R. Hess as described in Chapter 1.

In a branched vascular system of z generations of dichotomy, starting with z = 0 and progressing to z = 1, 2, 3, 4, . . . the size of each vessel segment must be smaller than that of its parent by a factor of $\sqrt[3]{1/2}$ which means that the radius in generation z is

$$r_z = r_0 \cdot \left(\sqrt[3]{1/2}\right)^z = r_0 \cdot 2^{-z/3}$$

which yields a linear plot on semilog scales, as in Figure 6.6.

More detailed recent analyses of the problem have shown that the exponent 3 is probably incorrect, that it should rather be 2.7. Based on the concepts of fractal geometry and of scaling, this revised coefficient has been shown to apply to arterial systems (Suwa et al. 1963) and also describes the changes in vessel diameter during growth (Kurz and Sandau 1997).

· · · ·

not have to "pay" for the air we breathe. But I think it is still justified to say that there is a cost to making airways larger in order to reduce their resistance and that this cost is related to the airway cross-section r^2. This is so because to make airways larger, the wall of the tubes must be made thicker, so the cost of making the tube increases proportional to r^2. Note that Murray in fact also thought that the cost of making the blood vessel wall of a given size was an important component that contributed to the "material" costs. In addition, space is restricted in the lung, so increasing the airway cross-section causes indirect costs that are again proportional to volume or to cross-section. For these reasons, the arguments outlined in Box 6.1 are generally also valid for the airways; what will change is the absolute value of the cost factor, but that does not affect the basic relationship between the size of the airways as we proceed along the airway tree.

Let us therefore postulate that if the airways are designed for efficient ventilation their diameters follow Murray's law. We therefore predict that airway diameters are reduced by a factor of $\sqrt[3]{1/2}$ with each branching, so that in generation z the diameter d(z) is

$$d(z) = d_0 \cdot 2^{-z/3}$$

where d_0 is the diameter at the origin, that is, of the trachea.

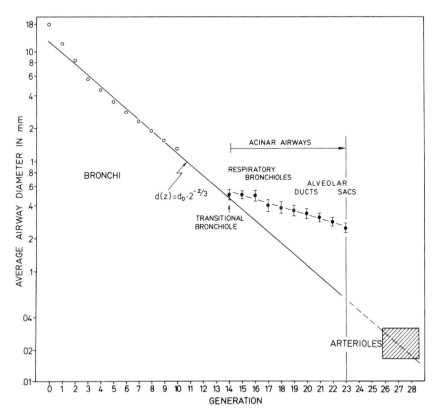

Figure 6.6 Semilogarithmic plot of average airway diameters against generation. From generation 0 to 14 (transitional bronchiole) the data are well described by Murray's law, as are the arteries, but the acinar airways deviate. (Data from Weibel and Gomez 1962 and Häfeli-Bleuer and Weibel 1988.)

This exponential function plots as a straight line on a semilog graph, and we predict that the airways of the human lung lie about this line if they are designed according to Murray's law. Figure 6.6 shows that this is the case for the mean diameter, averaged per generation, of the conducting airways of the first 14 generations, that is, down to the entrance airway into the acini, the terminal or transitional bronchioles (Figures 4.2 and 6.4). We conclude that in the conducting airways Murray's law is fulfilled. These airways are therefore designed for maximal efficiency of ventilation.

It is interesting to note that Murray's law can also be shown to apply for the pulmonary arteries. The major branches of the pulmonary arter-

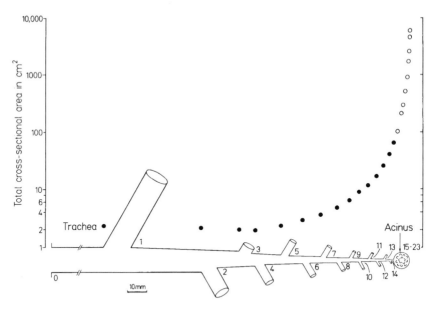

Figure 6.7 The so-called typical path model of airways drawn to scale along the abscissa, with increases of total airway cross-sectional area shown on a logarithmic scale. Conducting airways are represented as full dots, acinar airways as open circles. (From Weibel 1997.)

ies have the same dimensions as their accompanying airways (Figure 6.3), but they branch over more generations than the airways at the periphery. Notice in Figure 6.6 that the diameter range of the arterioles, the terminal branches of the arteries, falls exactly on the prediction line in generation 28, which is the average generation number at which we estimate arteries reach their terminal branches.

It is very evident from Figure 6.6 that Murray's law applies only to conducting airways and not to the approximately nine generations of airways in the acinus, the actual gas exchange unit. Airways of generations 15–23 carry alveoli in their wall (see Figures 4.2 and 6.4). But even if we measure only the central duct, we find that their diameter falls very slowly with each generation. Is this poor design? No, on the contrary. The conditions for Murray's law are not fulfilled in these peripheral airways. The presence of alveoli in the walls of acinar airways means that the air in the duct is not merely in transit in order to deliver oxygen to some distant periphery, but must also supply oxygen to its own alveoli; this occurs by radial diffusion of oxygen through the air phase. Air flow velocity has also become very low, because of the increase of the total cross-section of the airways (Figure 6.7). Thus even in the longitudinal direction the diffusion of oxygen in air is just as fast and becomes the major mechanism for "alveolar ventilation" within the acinus, which ensures the replenishment of O_2 in the air near the gas exchange surface. The resistance to diffusion is drastically reduced if the airway cross-section is large. We therefore conclude that it is of functional advantage not to reduce the airway diameters in the acinus according to Murray's law, but to maintain their cross-sectional area as large as possible. Murray's law applies to the mass transport of blood or air in a tube system, whereas the laws for maximal efficiency of oxygen diffusion in the gas phase are clearly different. The design of the airways in the lung is apparently such that it complies with the laws of maximal efficiency applying to the specific functional processes.

Are Airways Designed as Fractal Trees?

There is another way of looking at airway design. Consider that the branching pattern of airways is similar at all levels. In Figure 6.5, in comparison with Figures 4.1 and 4.2, it is quite striking that the pattern of branching, the shape of the bifurcations at each node, is very similar whether we look at large bronchi, medium-sized airways, or at the terminal bronchioles that lead into the acinus. In other words, regardless of scale the units of the airway tree resemble one another. This is an example of what is called scale-invariant self-similarity. It is the result of lung development that progresses systematically by the branching of the terminal tube combined with the proportional growth of the airways (Figure 6.1).

Scale-invariant self-similarity is a widespread biological phenomenon. We find examples mainly in the world of plants, such as in the cauliflower, where each lobe is a small replica of the whole cauliflower, and this can be carried down to very small units. It is indeed worth the experience to dissect a cauliflower with a morphologist's eye in order to discover the pattern of scale-invariant self-similarity. Other examples are trees such as beeches or oaks, whose branching patterns are clearly self-similar: a young sapling has the same form as the terminal twigs of its parent tree.

Self-similarity is the basic feature underlying a new geometry of nature, called fractal geometry, as introduced by Benoit Mandelbrot (1977, 1982). Is the lung designed like a fractal tree? If so, this would have interesting consequences for our interpretation of the relation between form and function in the lung.

The fractal model of the "Koch tree," constructed by Mandelbrot and shown in Figure 6.8, is often associated with the airway tree, and the question is whether it provides a valid model. What is the structure of this "Koch tree"? Based on the principle of constructing fractal models, the "Koch tree" is built from a generator which in this case is a T-shaped element with fixed proportions in the length and diameter of the arms. It is evident from Figure 6.8 that each daughter branch becomes a parent for the next branching and that this continues throughout to the end. In fact, there is no end to the "Koch tree"; the graph has simply been broken off after 12 generations. A primary virtue of this tree is that it fills, with its tips, the space homogeneously, and indeed very densely. But one of its most interesting aspects is that the distance from each end tip to the origin is equal for all tips irrespective of whether they are close to the origin or at the outermost corner, as can be verified in Figure 6.8. If the lung can be conceived as a fractal tree for which this "Koch tree" is an ideal model, then the prediction is that the pathway length for ventilation from the trachea to all the end tips of the airway tree is approximately equal by basic design.

Is the airway system a fractal tree? We can proceed to two tests. The first is to determine whether the principle of scale-invariant self-similarity can be found and quantified. Are the proportions between length and diameter maintained between the airway segments of one generation and from generation to generation? The average length and diameter of airways vary considerably. But when such measurements are averaged for each generation, we find that the proportions are indeed constant. The length-to-diameter ratio is invariant at 3.25 and the branching ratios, that is, the factors by which the dimensions are reduced with each successive generation, are 0.86 for diameter and 0.62 for length, irrespective of the generation. These findings support the

Airways and Blood Vessels · · · 147

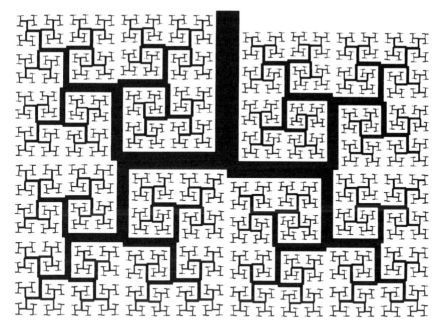

Figure 6.8 Fractal tree model simulates space-filling branching of airway tree. (From Mandelbrot 1983.)

notion that the airway tree has the basic properties of self-similarity, in spite of irregularities.

A second test for the fractal nature of the airway tree involves looking at the tree as a whole. Self-similarity or the scale-invariant constancy of proportions leads to the prediction that the diameter of airways should be a power law function of the scale, and if the scale changes with each generation it should be a power law function of the generation. The test is to plot the log d against log z, where a straight line relationship should result. If the data plotted in Figure 6.6 are replotted on a double logarithmic scale (Figure 6.9), we see that the data swing around the double log regression line. From this type of analysis, performed on several species, Bruce West and his colleagues concluded that the airway tree has the basic properties of a fractal tree and that the deviation of the data points from the straight line relationship must be interpreted as a harmonic variation; it may well have to do

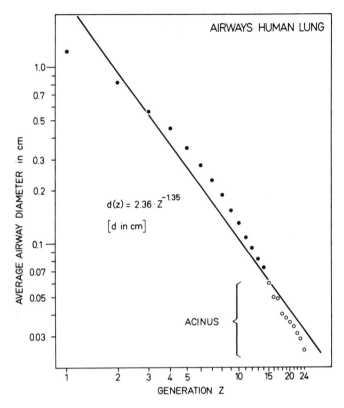

Figure 6.9 Figure 6.6's data for airway diameters, plotted as a power law against the generation to test for the fractal nature of airways. (From Weibel 1991.)

with the irregularities of branching imposed by the fact that the shape of the lung is determined from outside, by the shape of the space available in the chest.

We conclude that the airway branching approximates the ideal model for homogeneous ventilation, as exemplified by the "Koch tree." At the same time, it is quite inconceivable how a space-filling tree that needs to ventilate a very large number of units efficiently could be constructed much differently. The fractal tree principle appears to be a natural solution to a difficult problem, one that has been employed over and over again in nature with different variants according to specific needs.

In this regard it may be interesting to ask whether Murray's law bears any relation to the concept of fractal trees. This is definitely the case. Benoit Mandelbrot has discussed this law in terms of fractal geometry, and has noted that in a space-filling tree the diameter exponent of the sequence of branches should be close to 3 or a little smaller and that accordingly the diameter branching ratio should be about 2 to the power 1/3 or 1/2.7, as observed.

Conclusion

At the beginning of this chapter I outlined the problem of shaping the very large gas exchange surface of the lung in such a way that it can be accommodated within the chest cavity and efficient and homogeneous ventilation and perfusion become possible. The formation, during the fetal period, of parallel and interdigitating trees of airways and blood vessels allows this to happen in an orderly fashion and results in the parallel arrangement of three fractal trees. Because the gas exchange surface is built on the wall of the most peripheral airway generations, the alveolar surface itself has the properties of a fractal surface—where the surface complexity is the result of systematic folding, with smaller and smaller units being added toward the periphery.

Two points are noteworthy from the perspective of the importance of design for functional performance. As a consequence of arranging a fractal surface on a fractal tree, the distances from the origin of airways and blood vessels in the center of the chest to the gas exchange units vary over a relatively small range. This design ensures to an important degree that ventilation and perfusion of all units are homogeneous. Because the airway and vascular trees are parallel, this design also favors the match between ventilation and perfusion.

The design principle of fractal trees also ensures that the airways and blood vessels are designed for efficient air and blood flow, in that the dimensions of the tubes are systematically reduced from center to periphery for the greatest economy of operation. On the whole, therefore, the principle of economic design adjusted to the functional needs of the organ is satisfied in the structures supporting ventilation and perfusion of the lung. This design indeed lends strong support to the principle of symmorphosis.

7: The Pathway for Oxygen: From Lung to Mitochondria

In the preceding chapters we have looked at muscle cells, at capillaries, and at the lung, asking how structure is related to function in these particular instances. In this chapter I examine some of the observations made, with the goal of working out the rules that govern the integrated design of the entire pathway for oxygen, from the lung to the mitochondria, to see whether the principle of integration of function by design is at play. In such a functional system different organs, which are lined up like a chain, must work in concert, and this means that their functions must all be coadjusted to the task they share. The question now is whether the process of integration of functions goes along with the quantitative coadjustments of the structures involved—in other words, whether all links of the chain are made commensurate to the overall function of the system: to produce oxidative energy to support the work of the muscles.

We have looked in detail at mitochondria, capillaries, and the lung because they are all related to one basic function: oxidative energy production. As such they are part of the respiratory system, or the pathway for O_2. Mitochondria and the lung span the system from entry to exit, but there is a missing link: the circulation of the blood, which is the chief integrative feature that links the organ of O_2 uptake to that of O_2 consumption. It is driven by the heart, and it is well known that the heart readily adjusts to altered functional demands, as discussed in Chapter 1. As we begin to exercise, the heart pumps blood at a higher frequency, thus increasing blood flow between the lung and the muscle. Increased oxygen supply to muscles must be matched by increased uptake in the lung, and this, in turn, requires ventilation of the lung to be augmented by taking deeper breaths at a higher frequency. These are all functional regulations that lead to coadjustment of the O_2 transport functions at all steps of the respiratory system, with the result that O_2

uptake at the mouth is precisely matched to O_2 consumption in the muscle cells. Upon initiating muscle work these regulatory events are rapid, occurring in seconds, so that the overall function of O_2 transport is adjusted to a new steady state in a minute or two.

Each of these functional processes, however, can be upregulated only up to a well-defined limit, and as a consequence overall O_2 consumption, transport, and uptake is limited to $\dot{V}_{O_2}$max, the global measure of the functional capacity of the system (Figure 2.3). One of the central questions I wish to address in this chapter is whether this overall limitation of O_2 supply through the respiratory system is due to a single limiting step, one bottleneck along the pathway, or whether the functional capacities of the different steps in lung, heart, circulation and muscle are coadjusted to one another, as we would predict from symmorphosis.

My approach here is somewhat unusual, for several reasons. First, to take an integral look at an entire functional system and its design will force us to deviate, in part, from the reductionistic paradigm of contemporary science. I shall certainly be reductionistic when dissecting the various elements of the pathway, but this will not be enough. We will have to find ways to reintegrate the detailed findings on a broad scheme. I will ask whether the pathway for oxygen is "well designed," using as a criterion the principles of adaptation, integration, and economy—ultimately of symmorphosis—introduced in Chapter 1. One of the main reasons I suspect that the body's elements are optimally designed is the enormous inner complexity of the organism—and looking at a functional system as a whole is a first step toward allowing that complexity to enter our considerations. The body must make sure that all the functions that make up the complexity of the organism are adequately served, and all this must be accommodated within the limited space of the body, circumstances suggesting that economic design must prevail.

A further strong argument that "good design" may be important is the great and spectacular diversity of animals, the wealth of the biosphere. In previous chapters I discussed the effect of adaptation on the design of muscle and the lung to different lifestyles, using dogs, goats, and pronghorns as examples. The horse and the cow are a similar adaptive pair in another size class. But the most striking variation is the body size of mammals, which range from the smallest species, the

152 · · · Symmorphosis

Figure 7.1 The smallest mammal, the Etruscan shrew *(Suncus etruscus)*, which is from the Mediterranean region and weighs about 2 grams. (From Weibel 1979.)

Etruscan shrew *Suncus etruscus* which weighs just 2 grams (Figure 7.1), to elephants and whales weighing several tons. There is a wealth of literature on the effect of scale on form and function, on what is called allometry. We too will make use of this variation because it is well known that small animals have a higher metabolic rate per unit body mass than large animals, so that the oxygen needs of the body will be larger.

Testing the Hypothesis of Symmorphosis

To approach the question of whether form and function are coadjusted in an entire functional system requires that we study a system of sufficient complexity where the overall functional requirements vary to a significant and measurable degree.

For this purpose I first choose the pathway for oxygen from the lung to the mitochondria (Figure 7.2), before adding fuel supply in the next

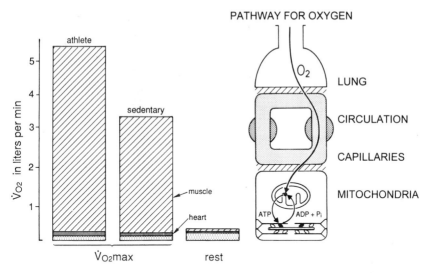

Figure 7.2 Variation in oxygen consumption in humans from rest to exercise in a sedentary and in an athletic individual, with distribution of oxygen consumption to muscle and heart. (From Weibel, Taylor, and Hoppeler 1992.)

chapter. Why is the pathway for O_2 a good test case for our general hypothesis? There are four main reasons:

1. This system serves one dominant function, namely the supply of oxygen from ambient air, and in mammals it is the only organ to serve this function;
2. The overall function has a measurable limit, $\dot{V}_{O_2}$max, inducible by activation of a single organ system, the locomotor musculature, which in exercise consumes over 90% of the oxygen taken up (Figure 7.2);
3. This limit varies to a significant degree among individuals and species;
4. The pathway is made up of a sequence of linked and obligatory structures for which engineering models can be devised regarding the way in which their parameters affect oxygen flow, and—an important aspect—there is no "escape pathway" so that the system compares with a simple linear chain model that is easy to treat.

How should we approach a test of the hypothesis of symmorphosis in such a system? We note that the functional performance of the system, in our case the flow rate of oxygen from the lung to the mitochondria, $\dot{V}_{O_2}$max, is variable in two ways (Figure 7.2). First, it is upregulated during work in proportion to the instantaneous needs for oxidative phosphorylation, as outlined above and in Chapter 2 (Figure 2.3). This rapid process, with response times of seconds to minutes, results in an increase in oxygen consumption between rest and maximal work by a factor of about 10 or more. This regulation involves functional parameters, such as heartbeat frequency or respiratory frequency, as well as adjustment of the enzyme rates of the respiratory chain, and so on. The functional capacity of the system sets the upper limit for these regulatory processes. This capacity is also variable, for example as a result of exercise training, or because of adaptive traits.

Since I postulate that the functional capacities are, to an important degree, determined by quantitative design properties, it follows that symmorphosis must be tested with respect to the functional capacities of each level in relation to the overall capacity of the system, in this case $\dot{V}_{O_2}$max. In other words, the relationship between form and function is felt when the limit of functional performance is approached or reached.

The Strategy: Exploiting Comparative Physiology

I assume that $\dot{V}_{O_2}$max estimates the limiting capacity of the entire respiratory system. This limitation of performance can either be due to a bottleneck at one point of the system—generally thought to be at the heart—or represents the matched capacity of all parts of the system. This distinction is crucial, and it determines the strategy we use.

Our postulate runs as follows: if there is a single bottleneck along the pathway and all other capacities are excessive in an arbitrary way, then only that specific bottleneck becomes precisely adjusted to $\dot{V}_{O_2}$max when overall aerobic capacity varies. Conversely, if the serial capacities are matched and coadjusted to one another—as symmorphosis predicts—then they would all be varied in parallel with the variation in $\dot{V}_{O_2}$max. This does not preclude the possibility that, even in a well-coadjusted pathway, one particular step will critically limit the functional performance at any time, but it may be one or the other, the heart or the lung, for example, depending on the prevailing conditions.

The variation in $\dot{V}_{O_2}$max is evidently crucial, and its precise quantitative relation to the complex chain of transport steps is not easy to work out. For this reason my team and that of C. Richard Taylor chose a strategy that considers two different causes of $\dot{V}_{O_2}$max variation: allometric and adaptive variation.

The largest variation in O_2 needs is observed in *allometric variation*. It has long been known that small animals have a relatively higher metabolic rate than large animals per unit body mass. This was first thought to be related to the differences in the surface-to-volume ratio of the body, which is larger in small animals because it varies with body mass to the $2/3$ power. Then Max Kleiber established, in 1961, that standard $\dot{V}_{O_2}$max increases with body mass to the $3/4$ power. In terms of specific metabolic rate, this means that O_2 consumption per unit body mass falls with $M_b^{-0.25}$ (Figure 7.3). As a consequence, a mouse of 30 g has a standard metabolic rate per unit body mass ten times higher than that of a 300 kg cow. Interestingly, it is found that $\dot{V}_{O_2}$max shows a similar variation with body mass, since the aerobic scope—the capacity to increase oxidative metabolism in exercise—is about tenfold in animals of all size classes (Figure 7.3). As a result the allometric equation for $\dot{V}_{O_2}$max is nearly parallel to Kleiber's line, namely

$$\dot{V}_{O_2}\text{max}/M_b = 1.94 \cdot M_b^{-0.21} \tag{7.1}$$

as compared with Kleiber's equation

$$\dot{V}_{O_2}\text{max}/M_b = 0.188 \cdot M_b^{-0.25} \tag{7.2}$$

$\dot{V}_{O_2}$max can also, however, vary among animals of the same size class: horses have a standard $\dot{V}_{O_2}$max similar to that of cows, but in running they achieve a $\dot{V}_{O_2}$max that is 2.5 times higher because of their athletic prowess (Figure 7.3). The same is true if we compare ponies with calves, dogs with goats, or foxes with other animals weighing a few kg. In the size class of ~30 kg we also find the superathlete pronghorn (Figure 1.1), which achieves a $\dot{V}_{O_2}$max that is 2 times higher than in the dog and 5 times higher than in the goat, as discussed in Chapter 2. The higher aerobic scope of athletic species, which is typically around 30-fold, causes their $\dot{V}_{O_2}$max to be higher than that of sedentary species of the same body mass by a factor of about 2.5. In top athletes, such as pronghorns or thoroughbred racehorses, the aerobic scope may rise to

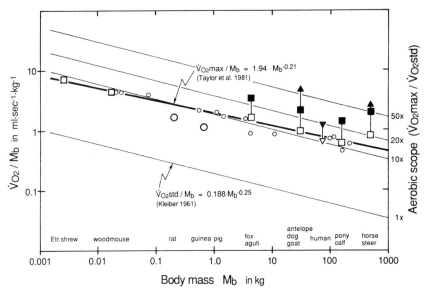

Figure 7.3 Allometric plot of mass-specific O_2 consumption, $\dot{V}_{O_2}/M_b$, showing Kleiber's curve (1961) for standard (basal) and isopleths for theoretical aerobic scope (fine lines), the allometric plot for $\dot{V}_{O_2}max/M_b$ (open circles, heavy line; Taylor et al. 1981). Adaptive pairs are shown as open and closed squares for sedentary and athletic species, respectively (Taylor et al. 1987; Jones et al. 1989; Longworth et al. 1989); closed upright triangles show the highest values measured in these size classes on thoroughbred racehorses (Evans and Rose 1987) and pronghorns (Lindstedt et al. 1991). Human $\dot{V}_{O_2}max$ data are for sedentary people and highly trained athletes (open and closed triangles, respectively). (From Weibel et al. 1992.)

about 50 or 60 times standard $\dot{V}_{O_2}max$ (Figure 7.3). This capacity to reach different levels of maximal O_2 consumption is called *adaptive variation*. It is due to constraints very different from those operative in allometric variation, so that we may expect the different parts of the pathway for oxygen to be differently modified when they must support a higher $\dot{V}_{O_2}max$ owing to one or the other of these types of variation.

To compare athletic and sedentary species of the same body size is relatively easy; although it is quite demanding to run horses and cows on a treadmill, such studies correspond to well-controlled laboratory experiments. This is not the case when we try to work out the principles underlying allometric variation. Here we wish to explore how the design of the respiratory system varies in the whole range of mammals from the shrew and the mouse to the cow and the elephant. The conditions of life are clearly very different for the different size classes. Whereas smaller mammals are still found in their—more or less—natural habitat, this is not the case for animals larger than a deer in many parts of the world. This is why, at the suggestion of C. Richard Taylor, our colleagues and I conducted the crucial parts of an allometric study, on which I report here, in fieldwork in Kenya in 1977. Here animals of all sizes still live in their natural habitats, and we could collect mammals that lived under similar conditions, from the small mongoose of about 500 g up to the Eland antelope that weighed 150 kg, thus spanning three orders of magnitude. This sounds like an ideal situation, but it was, in actual fact, a rather difficult and demanding study, and it could only be brought to fruition through the active help of our African friends and colleagues, especially Geoffrey Maloyi and Wangari Muta Matai and their collaborators at the University of Nairobi. We had also wanted to study water buffaloes and elephants, which would have extended the scale of body size by at least one order of magnitude, but that proved technically—and politically—impossible at the time. Larger animals, such as horses and cows of 500 kg body mass, were added to this allometric study later, as were the very small species such as rats, mice, and shrews, including the Etruscan shrew (Figure 7.1). In all, the animals studied covered over five orders of magnitude in body mass.

The tactical approach was as follows. First an extensive physiological study was performed, with the animals running on a treadmill of appropriate size—the one used for large African mammals was built in Boston and transported to Kenya by ship. And the physiological study on horses and steers was done at Uppsala in Sweden, where Arne Lindholm could offer not only excellent equipment for working with these large and powerful animals but also a congenial and generous partnership—at least as important a prerequisite for successful work as good hardware. After completion of these studies the animals were sacrificed, and the lung, the heart, and all the locomotor muscles were sampled for a later morphometric study by light and electron microscopy.

These studies lasted for nearly fifteen years and involved a large number of active collaborators. They are listed in the Preface, not only to thank them but to document that such studies cannot be done as one-

or two-man shows. But consistent leadership is also essential. The physiological studies were done under the direction of C. Richard Taylor, in his laboratories at Harvard's Concord Field Station as well as in Nairobi and Uppsala with the assistance of Richard Karas, James Jones, Arne Lindholm, and many others. The morphometric studies were performed in our laboratories in Berne in Switzerland with Hans Hoppeler and Peter Gehr and many collaborators, assisted also by specialists in morphometric and stereological methods, primarily Luis Cruz-Orive. Finally, the data were pulled together and synthesized in a collaborative effort between the two research groups at Harvard and in Berne.

The question we now ask is this: since small and athletic animals must move more oxygen through the respiratory system, is the design of the different steps through which O_2 is transported (Figure 7.2) adjusted at all levels, or is the adjustment limited to one step that would represent a bottleneck?

An important consideration in examining these two types of variation is that changes in the size of structures that are not related to the varying needs for oxygen consumption will be of a different nature. This should help us in sorting out those effects that can be considered true adaptations to higher needs in O_2 transfer from those related to other conditions.

The Model and Predictions

The last requirement before we can proceed with the analysis is to set up a quantitative model of the respiratory system that defines the role of the functional and the structural parameters at all steps. The model of Figure 7.4 describes the oxygen flow rates in five steps: into the lung, from the air to the blood, through the circulation, from the microcirculation into the muscle cells, and finally through the respiratory chain in mitochondria resulting in oxygen consumption. It is important to note that under steady-state conditions the O_2 flow rates through the serial steps must be the same, because no O_2 disappears along the way, and that they can therefore be estimated by measuring oxygen uptake at the mouth and nose. Oxygen uptake therefore equals oxygen consumption in the mitochondria. This condition is naturally fulfilled because there is no other way of getting O_2 into the body except through the lung.

The basic driving force for oxygen flow through the system is a cas-

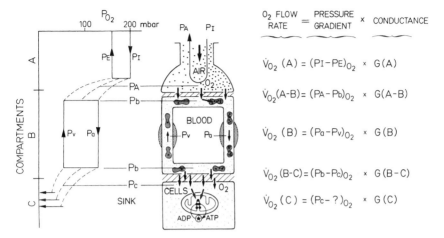

Figure 7.4 Model of the pathway for oxygen from inspired air to the respiratory chain in the mitochondria. To the left is the partial O_2 pressure cascade that serves as driving force for O_2 flow; to the right, the set of equations that relates O_2 flow rate to the pressure differences and the conductances at each level. (Modified after Weibel and Taylor 1981.)

cade of oxygen partial pressures that fall from inspired P_{O_2} down to near zero in the mitochondria (Figure 7.4). The graph also indicates that there are two convective steps, ventilation of the lung and circulation of the blood, and two diffusive steps, from air to blood in the lung and from capillary blood to mitochondria in the muscle. In the mitochondria, finally, O_2 disappears as it is consumed and reduced to water in oxidative phosphorylation, an enzyme-catalyzed process, as we saw in Chapter 2.

The oxygen flow rates at these different levels can be expressed as the product of a pressure difference (resulting from the pressure cascade) and a conductance (Figure 7.4). It is very convenient for our purposes that nearly all the structural variables are contained in the conductance terms. One such term has been worked out in Chapter 4, namely the diffusing capacity of the lung, here expressed as $G(A - B) = D_{L_{O_2}}$. These conductances are of three types: for ventilation and circulation they are convective conductances; in the pulmonary and tissue gas exchangers they are diffusive conductances; in the mitochondria the conductance depends on enzyme kinetics.

In order to be useful for our purposes this model must be refined by clearly separating the functional and structural variables. Such a refined model is described in Box 7.1. Just to indicate how this model differs from the one in Figure 7.4, consider the third term, describing oxygen transport by the circulation. This is generally described by the Fick equation as the product of cardiac output $\dot{Q}$ multiplied by the oxygen concentration difference between arterial and mixed venous blood:

$$\dot{V}_{O_2} = \dot{Q} \cdot (Ca_{O_2} - C\bar{v}_{O_2}) \tag{7.3}$$

Each of these terms contains a functional as well as a design parameter. Cardiac output is the product of heart frequency times stroke volume V_s as briefly described in Chapter 1: $\dot{Q} = f_H \cdot V_s$, where heart frequency is clearly a functional variable which is rapidly increased when the needs for oxygen transport demand higher blood flow; in contrast, stroke volume cannot be increased when blood flow increases since it is directly related to the size of the heart chambers and is therefore a morphometric design parameter. The oxygen concentration in the blood depends first on the hematocrit or the relative volume of erythrocytes, $V_V(ec)$, a morphometric parameter of blood, and then on the relative saturation of the hemoglobin with oxygen. This saturation is a function of the prevailing oxygen partial pressures, Pa_{O_2} and $P\bar{v}_{O_2}$ in arterial and venous blood, respectively, multiplied by the respective oxygen capacitance of the blood, σ_a and $\sigma_{\bar{v}}$. This capacitance depends on the O_2-hemoglobin dissociation curve, which is a nonlinear function. Thus we have $Ca_{O_2} = \sigma_a \cdot Pa_{O_2}$. With these substitutions we can now sort out equation 7.3 to separate functional and design parameters:

$$\dot{V}_{O_2} = \underbrace{(\sigma_a \cdot Pa_{O_2} - \sigma_{\bar{v}} \cdot P\bar{v}_{O_2})}_{\text{functional}} \cdot f_H * \underbrace{V_s \cdot V_v(ec)}_{\text{design}} \tag{7.4}$$

where all the functional variables are to the left of the large multiplication asterisk and the morphometric variables are to the right. In the model shown in Box 7.1 this type of sorting out of the functional and structural variables is done for all four levels from the lung to the mitochondria.

On the basis of the hypothesis of symmorphosis we now postulate

> **Box 7.1 Model of the respiratory system separating functional and design parameters**
>
> | | Function | × Design |
> |---|---|---|
> | *Pulmonary gas exchanger* | | |
> | $\dot{V}_{O_2} =$ | $(P_{A_{O_2}} - P_{b_{O_2}})\{t_c, \theta_{O_2}\}$ | × **$D_{L_{O_2}}$** $\{S(A), S(c), V(c), V_V(ec), \tau_{hb}\}$ |
> | *Circulation* | | |
> | $\dot{V}_{O_2} =$ | $(\sigma_a \cdot P_{A_{O_2}} - \sigma_v \cdot P_{v_{O_2}}) \cdot \mathbf{f_H}$ | × **V_s** $\{V(LV)\} \cdot \mathbf{V_V(ec)}$ |
> | *Microvasculature* | | |
> | $\dot{V}_{O_2} =$ | $(P_{b_{O_2}} - P_{c_{O_2}})\{t_c, \theta_{O_2}\}$ | × **$D_{L_{O_2}}$** $\{S(c), V(c), V_V(ec), \delta(c,m)\}$ |
> | *Mitochondria* | | |
> | $\dot{V}_{O_2} =$ | $\dot{v}_{O_2}\{\dot{m}_{ATP}\}$ | × **V(mi)** $\{S_V(imi, m)\}$ |
>
> The O_2 flow rate $\dot{V}_{O_2}$ is expressed as the product of functional and design parameters shown in boldface; parameters that affect the factors are shown in italics and placed in braces { }. The functional parameters include: O_2 partial pressures (P_{O_2}), coefficients of "hematocrit-specific" O_2 capacitance (σ) which depend on O_2-hemoglobin dissociation, O_2 binding rate (θ), heart frequency (f_H), capillary transit time (t_c), and mitochondrial O_2 consumption rate as function of ATP flux ($\dot{v}_{O_2}\{\dot{m}_{ATP}\}$). Design parameters include: diffusion conductances (D) of lung and tissue gas exchangers, which depend on alveolar and capillary exchange surface areas [S(A), S(c)], capillary volumes [V(c)], hematocrit [$V_V(ec)$], harmonic mean barrier thickness (τ_{hb}), capillary-mitochondrial diffusion distance [$\delta(c - mi)$], and mitochondrial volume [V(mi)] with inner membrane surface density [$S_V(imi, mi)$]. (Modified after Weibel, Taylor, and Hoppeler 1991.)
>
> · · · ·

that the design parameters of the various steps are coadjusted to the maximal functional demand set by $\dot{V}_{O_2}$max. From this we now predict that when $\dot{V}_{O_2}$max varies, be it by allometric or by adaptive variation, the structural parameters vary in parallel with functional capacity. To put it another way, with varying $\dot{V}_{O_2}$max the ratios of design parameters to functional capacity are invariant, which may be written as follows:

[design parameters]/{functional capacity} = invariant

as was briefly introduced in Chapter 2.

We are now ready to test the hypothesis of symmorphosis by exploiting the large differences that nature offers through allometric and adaptive variation as the two key approaches of comparative physiology. We proceed as follows: we shall estimate the design parameters at each level, beginning at the mitochondria and moving up to the lung, seeking invariant ratios between design parameters and functional capacity under the two modes of variation, allometric and adaptive, but we shall accept them as invariant only if they are such in both modes.

This analysis will need a very large data set from which we wish to extract some general information. These data will be presented in two forms. (1) For adaptive variation I shall present a set of tables (Tables 7.1–7.4) with the actual data for three pairs of athletic/sedentary species: dog/goat, pony/calf, horse/steer; the data are presented according to the parameters that define the structure-function relation (Box 7.1). For each parameter I have also worked out the ratios of athletic to sedentary species, to be used in the interpretation of the results. (2) For allometric variation I present a set of graphs (Figures 7.5–7.8) with the data points for "normal" animals (open symbols) plotted on double logarithmic scales against body mass, in which the power law regression is a linear curve with the slope corresponding to the exponent of the power law. In these graphs the data points for athletic animals are plotted as black dots. For the interpretation of this large data set, the athletic/sedentary ratio of adaptive variation and the slopes of the power law regression lines of allometric variation will be most important.

Testing the Respiratory System for Symmorphosis

Do Muscle Mitochondria Set the Aerobic Capacity?

In the model of Box 7.1, the last line postulates that the functional parameter characterizing oxygen flow into the mitochondria is the rate of oxygen consumption of a unit mitochondrial volume which then, by necessity, leaves mitochondrial volume as the main design parameter.

The Pathway for Oxygen · · · 163

The arguments that justify using a constant specific rate of mitochondrial oxygen consumption were discussed extensively in Chapter 2: (1) under fully activated conditions the ratio between oxygen flux and ATP turnover is the same for all mitochondria; (2) the surface density of inner mitochondrial membrane was found to be invariant between all muscle cells in mammals; (3) in vitro estimates of oxygen consumption per inner mitochondrial membrane gave the same values for muscles of different activity potential.

These assumptions are also largely correct for heart mitochondria, but not for "other" mitochondria. For example, mitochondria in liver cells have a considerably lower surface density of inner mitochondrial membranes. It is important to note, however, that in heavy exercise 90% of oxygen consumption occurs in the skeletal muscles, about 5%–7% in the heart muscle, and accordingly less than 5% in other cells (Figure 7.2). Taking all the muscle mitochondria together, in the athletic species about 4% of the mitochondria are in heart muscle as compared with about 7% in the sedentary species. This correlates quite nicely with the estimate of oxygen consumption by the heart, which is about 4% in athletes and 6% in sedentary species. For all these reasons I will base this analysis on the comparison of locomotory muscle mitochondria with $\dot{V}_{O_2}$max and neglect all other mitochondria.

Following our strategy, we first compare athletic with sedentary species, expanding on the analysis done in Chapter 2. Table 7.1 shows the relationship between $\dot{V}_{O_2}$max and mitochondrial volume for three species pairs in which one is an athletic and the other a more sedentary animal. The relationships are quite consistent: the athletic species have a mitochondrial volume that matches their maximal oxygen consumption rate. This confirms the finding of Chapter 2 that for every ml of oxygen consumed at $\dot{V}_{O_2}$max, there are 15 ml of mitochondria in the skeletal muscle cells, or, conversely, that each ml of mitochondria consumes about 5 ml O_2 per minute.

Turning now to allometric variation, we have data available from mammals ranging from a mouse weighing 20 g to the 300 kg cow. Figure 7.5 shows an allometric plot of $\dot{V}_{O_2}$max and of total muscle mitochondrial volume with log body mass as the abscissa. We clearly see that both the functional and the morphometric parameter increase with the same slope, that is, with the same power of M_b. Figure 7.5 also shows that the corresponding data points for four pairs of athletic

Table 7.1 Mitochondrial volume related to aerobic capacity

	$\dot{V}_{O_2}max/M_b$ $ml \cdot sec^{-1} \cdot kg^{-1}$	$V(mi)/M_b$ $ml \cdot kg^{-1}$	$[V(mi)]/\{\dot{V}_{O_2}max\}$ $ml \cdot ml^{-1} \cdot sec$
Dog	2.29	40.6	17.7
Goat	0.95	13.8	14.5
D/G	2.4*	2.9*	1.2
Pony	1.48	19.5	13.2
Calf	0.61	9.2	15.1
P/C	2.4*	2.13*	0.9
Horse	2.23	30.0	13.5
Steer	0.85	11.6	13.7
H/S	2.6*	2.6*	1.0
Ath./Sed.	2.5*	2.5*	1.0

Source: Data from Weibel, Taylor, and Hoppeler (1991).
*Significant ratio.

species are systematically higher for both parameters, as we expect from Table 7.1.

We thus conclude that the ratio $[V(mi)]/\{\dot{V}_{O_2}max\}$ is invariant with both types of variation of $\dot{V}_{O_2}max$. The hypothesis of symmorphosis is therefore supported at this level. As a consequence, when muscle cells need more oxidative phosphorylation to support higher work rates, they build more mitochondria of the same type, confirming, now on a broader scope, the findings of Chapter 2.

A cautionary remark is appropriate in that this strict invariance between mitochondrial volume and $\dot{V}_{O_2}max$ may only be true in the "normal range" of mammals in which mitochondria typically occupy 5%–20% of skeletal muscle volume. But in very fast and/or very small animals this mitochondrial content may not be enough to provide the high levels of ATP required, and mitochondrial density must be increased. The question then is whether there is a limit to the amount of muscle cell space that can be occupied by mitochondria. For example, in the hummingbird flight muscle and in the heart muscle of the Etruscan shrew, both animals weighing about 2 g and constituting the lower

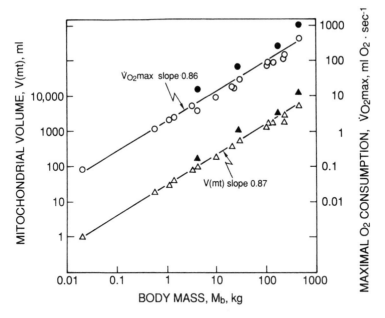

Figure 7.5 Allometric plot of maximal O_2 consumption and mitochondrial volume against body mass in mammals. Open symbols are sedentary "normal" species, black dots are athletic species. The regression curves are calculated for the "normal" species.

limit of body size in mammals and birds, mitochondrial volume density reaches about 50% of cell volume. Particularly in the hummingbird flight muscle it has been shown that mitochondria, in addition, change their nature in that the surface density of inner mitochondrial membranes is increased, as discussed in Chapter 2. This simply points to the fact that the morphometric parameter we have chosen reflects the functional capacity only because "normally" the density of inner mitochondrial membranes, the carriers of the respiratory chain enzymes, is invariant; it is however increased in cases of particularly high demand for oxidative phosphorylation, such as in the hummingbird flight muscle, and it is lower in reptilian muscles with their low energy demand. In this broader perspective it is the surface area of the inner mitochondrial membrane that forms an invariant ratio with functional aerobic capacity.

Are Capillaries Matched to Aerobic Capacity?

Chapter 3 presented evidence that the muscle capillaries are designed for the supply of oxygen to the muscle cells rather than for the supply of substrates when we compared dogs with goats. Is this also true on a broader scale, looking at both adaptive and allometric variation? Do we find the capillary volume to be proportional to $\dot{V}_{O_2}$max as we would predict from our general hypothesis? This prediction was indeed supported by our original study of the allometric relation between capillary volume and $\dot{V}_{O_2}$max, as shown in Figure 7.6: the two regression lines for $\dot{V}_{O_2}$max and V(c) run in parallel; the small difference in their slopes is not statistically significant. However, in this figure we also see that the capillary volume in athletic species is not increased in proportion to their larger $\dot{V}_{O_2}$max. Table 7.2 shows that the capillary volume is increased only by a factor of 1.7 in the athletic species as compared with 2.5 for $\dot{V}_{O_2}$max. In Chapter 2, I reported differences in the blood of dogs and goats which suggested that athletes tend to have a higher hematocrit than the sedentary species. Table 7.2 shows that this is true for all three species pairs. The horses included in this table were standardbred racehorses; it is known that in thoroughbreds the hematocrit can be even higher during strenuous exercise. Remember that Figure 2.11 presented evidence that the observed hematocrits of 30%–50% are in the range of "near-optimal," considering the cost of transporting oxygen. If in thoroughbred racehorses the hematocrit rises to 60%–70%, as reported, then the viscosity of the blood increases drastically, requiring high pressures to drive the blood through the lung and tissue, perhaps one reason why these horses tend to bleed into their lungs when they run.

On the basis of these observations, our original prediction that capillary volume is proportional to $\dot{V}_{O_2}$max is refuted. The capillary volume does not form an invariant ratio with $\dot{V}_{O_2}$max under all modes of variation. The results of the study of adaptive variation suggest that the morphometric parameter that varies in parallel with $\dot{V}_{O_2}$max is, in fact, the *capillary erythrocyte volume,* the product of V(c) and hematocrit $V_V(ec)$. Does this also apply to allometric variation? It does, because it turns out that hematocrit is invariant with body size, averaging between 30% and 40% in all species from the mouse to the cow; it only deviates from this optimum range (Figure 2.11) in athletic adaptation.

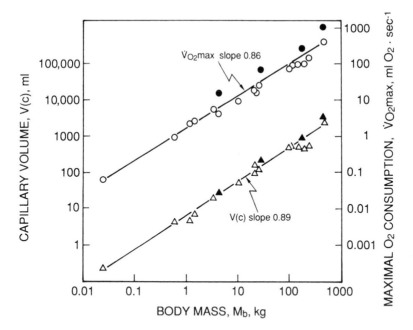

Figure 7.6 Allometric plot of maximal O_2 consumption and capillary volume against body mass in mammals. Symbols and regressions as in Figure 7.5.

We therefore find that, considering both adaptive and allometric variation, the invariant ratio with the functional capacity is [V(c) · V_V(ec)]/{$\dot{V}_{O_2}$max}.

We thus conclude that, in designing the gas exchanger of the muscle microvasculature, two structures are involved—the vessels and the blood—and that both adjust halfway to make a higher capacity for oxygen supply. The burden of adjustment is thus shared by the two independent structures, vessels and blood. This is indeed a very economical way of adjusting the system to higher demand.

Is the Heart Adjusted to Aerobic Capacity?

Oxygen is transported from the capillaries in the lung to the capillaries in muscle by the circulation of blood. This convective step (see Figure 7.4 and Box 7.1) is often considered the limiting step in oxygen transport during strenuous exercise. As already discussed, the transport of

Table 7.2 Capillaries and aerobic capacity

	$\dot{V}_{O_2}max/M_b$ ml·sec⁻¹·kg⁻¹	$V(c)/M_b$ ml·kg⁻¹	$V_V(ec)$ ml·ml⁻¹	$[V(c) \cdot V_V(ec)]/\{\dot{V}_{O_2}max\}$ ml·ml⁻¹·sec
Dog	2.29	8.2	0.50	1.79
Goat	0.95	4.5	0.30	1.42
D/G	2.4*	1.8*	1.68*	1.26
Pony	1.48	5.1	0.42	1.45
Calf	0.61	3.2	0.31	1.63
P/C	2.4*	1.6*	1.35*	0.89
Horse	2.23	8.3	0.55	2.05
Steer	0.85	5.3	0.40	2.49
H/S	2.6*	1.6*	1.4*	0.82
Ath./Sed.	2.5*	1.7*	1.5*	0.99

Source: Data from Weibel, Taylor, and Hoppeler (1991).
*Significant ratio.

oxygen by the circulation depends on the properties of the heart as a pump, measured by cardiac output, and of the blood as an oxygen carrier. We have seen that the Fick O_2 transport equation contains both functional and morphometric parameters (see Box 7.1, row 3). Cardiac output is the product of heart frequency f_H and stroke volume V_s, which is proportional to heart size.

What is varied in adaptive variation? Table 7.3 shows that in each of the three species pairs, maximal heart frequency is the same between the athletic and the sedentary animal. This is perhaps a surprising finding, because it is well known that human athletes have a lower *resting* heart frequency than normal people, and this is also the case in the athletic animals dog and horse as compared with goat and cow. What is a fixed value, however, is the *maximal rate* at which a heart can beat, and this maximal frequency is the same in both athletic and sedentary animals. As we shall see, this is true because maximal heart frequency is determined by body size. The entire difference in maximal cardiac output between athletic and sedentary species is therefore achieved by a higher stroke volume related to their larger hearts in the athletic species; again, it is well known that highly trained human athletes have larger hearts.

Table 7.3 Heart volume and aerobic capacity

	$\dot{V}_{O_2}max/M_b$ ml·sec^{-1}·kg^{-1}	$V_V(ec)$ ml·ml^{-1}	f_H min^{-1}	V_s/M_b ml·kg^{-1}	$[V_s \cdot V_V(ec)]/\{\dot{V}_{O_2}max/f_H\}$ ml·ml^{-1}
Dog	2.29	0.50	274	3.17	3.16
Goat	0.95	0.30	268	2.07	2.92
D/G	2.4*	1.68*	1.02	1.53*	1.08
Pony	1.48	0.42	215	2.50	2.54
Calf	0.61	0.31	213	1.78	3.21
P/C	2.4*	1.35*	1.02	1.40*	0.79
Horse	2.23	0.55	202	3.11	2.58
Steer	0.85	0.40	216	1.52	2.58
H/S	2.6*	1.4*	0.94	2.1*	1.00
Ath./Sed.	2.5*	1.5*	1.0	1.7*	0.96

Source: Data from Weibel, Taylor, and Hoppeler (1991).
*Significant ratio.

However, as seen from Table 7.3 the stroke volume of athletic species is larger than that of nonathletic species, but it is not proportional to $\dot{V}_{O_2}$max. This is not really necessary, because the athletic species have a higher hematocrit; thus one heartbeat pumps out a larger volume of erythrocytes even though the stroke volume is the same. The parameter that forms an invariant ratio to $\dot{V}_{O_2}$max is, in adaptive variation, the product of stroke volume times hematocrit, or the volume of erythrocytes ejected with each heartbeat. Here again we see that the effort of adjusting the circulatory transport function for oxygen to higher needs is shared between the blood and the heart, just as in microcirculation.

Let us look then at how O_2 transport by the circulation of blood varies with body size. Figure 7.7 shows that cardiac output increases in parallel with $\dot{V}_{O_2}$max with a slope which is not significantly different from that for $\dot{V}_{O_2}$max. And since there is no difference in hematocrit between small and large animals, this means that O_2 transport is proportional to needs at $\dot{V}_{O_2}$max. The design parameter of cardiac output, stroke volume, is directly proportional to heart weight, which in turn is a constant proportion of body mass over the size range of animals from mice to cows, namely about 0.58%, except for athletic species, where it

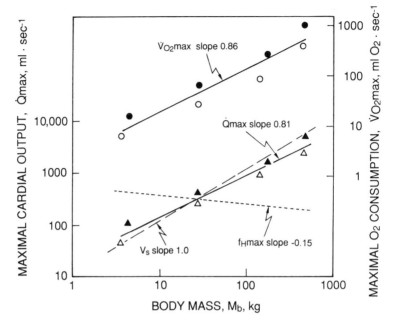

Figure 7.7 Allometric plot of maximal O$_2$ consumption and maximal cardiac output against body mass in mammals. Regressions of the two parameters, stroke volume V$_s$ and maximal heart frequency f$_H$, are shown as broken lines. Symbols and regressions as in Figure 7.5.

is some 50% larger. Accordingly, stroke volume increases linearly with body mass in sedentary animals, so that heart size and stroke volume are not adjusted to $\dot{V}_{O_2}$max in allometry. The difference is made up by maximal heart frequency, which falls with increasing body mass, with a slope of −0.15 (Figure 7.7).

It is indeed found that heart frequency varies with body mass M$_b$ in a very characteristic way in that small animals have higher frequencies than large ones, both at rest and at maximal exercise, but the slopes are different:

$$f_H \text{rest} = 209 \cdot M_b^{-0.26}$$

$$f_H \text{max} = 419 \cdot M_b^{-0.15}$$

An animal of 1 kg body mass thus has a resting frequency of 209 and can double this to serve the needs of maximal O_2 transport. In larger animals heart frequency can be augmented to a higher degree from rest to maximal exercise, for example by a factor of 3 in the human and even of 5 in a racehorse. In contrast small animals can increase their heart frequency to a much smaller extent, no more than 1.5 in the rat and 1.2 in the mouse.

In the context of our analysis the fact that maximal heart frequency is so tightly controlled by body size is most pertinent, particularly because we have observed that not even adaptation to very high performance, such as in the racehorse, can increase $f_H$max. It is plausible that the large heart of a cow cannot beat as fast as the small heart of a mouse — just as a cow cannot move its legs as fast as a mouse — but can we understand this in terms of the effect of design on function? What is behind this effect? We don't really know.

It may be useful to think of a parallel process such as maximal stride frequency, the maximal frequency at which animals can move their legs in locomotion. This too is size dependent and, in fact, shows a very similar allometric regression. It is a situation that has been better studied, but there is still dispute on the precise mechanism that sets the maximal frequency. In the example of limb movement the effect of size is easier to understand. Try a little test on yourself: with your hand steady, move your index finger up and down as fast as possible and estimate how many swings per second you can make; then do the same with your arm, and you will see that you can swing the larger limb only with considerably lower frequency. This is what happens in animals as they move their legs: large legs swing at a lower rate than small legs. Indeed, when a 30 g mouse runs fast each leg takes about five steps every second, but when a cow or a horse of 300 kg runs fast each leg takes only about one step every second. It is found that the maximal stride frequency, that is, the number of steps per second measured at the speed at which an animal changes from trot to gallop (cf. Figure 1.2), is an allometric function that falls with body mass to the power −0.14, interestingly the same exponent as that found for heart frequency. With stride frequency the limiting rate is largely explained by the length of the limbs, which swing like a pendulum, and the natural frequency of the pendulum is proportional to the square root of its length. This formula explains the exponent of −0.14, since leg length is approximately proportional to the cube root of body mass.

The size dependency of the maximal heartbeat frequency is less easy to explain, but the fact that it is described by the same power law exponent of −0.15 suggests that it is likewise related to the scaling of mechanical factors. The heartbeat is also a cyclical event which, from a mechanical point of view, involves two functional features: the heart ventricle must be filled with blood in diastole and emptied in systole so that the time interval between beats is adequate to allow the required amount of blood to flow into the chamber, and this takes longer in larger than in smaller hearts; this movement is achieved by a relaxation and expansion of the heart wall in diastole followed by a forceful contraction in systole. The active principle in this process is the heart muscle, which is made of short muscle cells assembled to form long fibers that are wrapped around the ventricle along so-called geodesic lines or meridians, similar to the twine in a ball of string. If we take a typical unit fiber, for example one that follows a meridian from the base of the heart to the apex, its maximal length is proportional to the cube root of heart volume in diastole, that is, when the chamber is maximally filled; and since heart volume is a constant fraction of body mass, it is proportional to body mass to the $1/3$ power.

This would suggest that a reasoning similar to that for stride frequency could lead to a similar explanation. Things are not as easy in this case, however, because the heart is not an oscillator like a pendulum; it is rather a piston pump with inflow and outflow valves and the heart muscle as the force generator. Such a pump is not limited by a maximal frequency in the sense of the oscillator but rather by the conditions of inflow and outflow, where turbulence caused by too high velocity of the blood must be avoided. At present we do not understand why maximal heart frequency is so tightly determined by body size, what the constraints are that act on the heart, and why the body size dependence of maximal heart frequency is different from that at rest.

Coming back to our attempt to analyze the structure-function relationships of the heart, let me briefly summarize briefly our findings: in adaptive variation the adjustment is achieved by varying stroke volume V_s and hematocrit $V_V(ec)$ with heart frequency constant; in allometry the adjustment is achieved by stroke volume and heart frequency f_H with hematocrit constant. Combining these two types of variation, we conclude that an invariant ratio must involve two structural and two functional variables, namely $[V_s \cdot V_V(ec)]/\{\dot{V}_{O_2}max/f_H\}$.

Can this be justified? I believe so, because, as we have seen, maximal heart frequency is an independent variable that establishes a ceiling for the rate at which the heart can pump blood. It is also a dominant factor in the sense that it is immutable, strictly determined by body mass. For this reason it determines the functional capacity of the heart as an oxygen transporter. If these arguments are accepted, then the hypothesis of symmorphosis is supported at the level of cardiovascular oxygen transport, but in this case three parameters are used to adjust the transport capacity: two structural variables, heart size and hematocrit, and a functional variable, heart frequency.

Is the Pulmonary Gas Exchanger Matched to Aerobic Capacity?

The lung's dominant function is to supply oxygen to the body in response to need, in exercise essentially in order to fuel muscle work through oxidative phosphorylation. It has therefore been an old notion that the lung's gas exchanger should be proportional to the oxygen consumption of the animal. The first quantitative investigation to suggest this was the classic allometric study of Tenney and Remmers (1963), who measured both O_2 consumption of the whole animal and alveolar surface area on dried lung specimens in a large series of mammals, ranging in body mass from 30-gram mice to 500-kilogram cows and including both terrestrial and marine mammals, thus allowing for very different lifestyles. This study concluded that the internal surface area of the lung is proportional to $\dot{V}_{O_2}$ when allometric variation of oxygen consumption is considered. A little later, our own study of the Japanese waltzing mice, described in Chapter 4 (Figure 4.8), showed that the lung's gas exchange surface is increased roughly in proportion to O_2 needs in the species with higher O_2 consumption. We thus came, in a case of adaptive variation, to the same conclusion as Tenney and Remmers for allometric variation.

More recent studies call into question certain earlier conclusions. First, we have learned that an unqualified estimate of oxygen consumption is not adequate; such comparisons must be based on $\dot{V}_{O_2}max$, the limiting condition of O_2 uptake. Second, we have seen in Chapter 4 that to estimate alveolar surface area alone does not provide an adequate description of the lung's gas exchange capacity because it disregards other important structural variables; we should rather base a solid structure-function relationship on estimates of the pulmonary diffusing capacity $D_{L_{O_2}}$, a complex parameter that is dominated by morphometric properties, one of which is the alveolar surface area (Box 4.1). To recapitulate briefly, the diffusing capacity is determined by two diffusion resistances in series: the membrane component, determined by the gas exchange surface divided by the harmonic mean barrier thickness, measured from the alveolar surface to the red blood cell surface; and the blood component, determined by the capillary volume and the hematocrit. The basic physiological coefficients of diffusion and of the reaction rate of blood with oxygen that enter the model calculations (Box 4.1) vary relatively little, but they can be used to weigh the morphometric variables in calculating a theoretical value for $D_{L_{O_2}}$. The advantage of this model (Box 7.1, row 1, and Box 4.1) is that it considers all major morphometric variables that characterize the pulmonary gas exchanger. This should therefore allow us to ask the question whether the pulmonary gas exchanger is made commensurate to the body's oxygen needs.

In that sense, our prediction from symmorphosis says that the pulmonary diffusing capacity $D_{L_{O_2}}$ should be proportional to maximal oxygen consumption $\dot{V}_{O_2}max$, or that the ratio $[D_{L_{O_2}}]/\{\dot{V}_{O_2}max\}$ is invariant under all modes of variation in oxygen needs, adaptive as well as allometric.

This hypothesis is not supported. When we compare athletic and sedentary species of approximately the same body mass (Table 7.4), we find that the athletic species have a gas exchange surface and a capillary volume that is about 1.3 times larger than in the sedentary species, which would make the membrane part of the diffusing capacity somewhat larger. An important difference lies, here again, in the hematocrit (Table 7.3), which is 1.5 times greater in the athletic species. The first effect of this is that the higher concentration of red blood cells causes the plasma layer that separates them from the tissue barrier to be thinner, which contributes further to increasing the membrane diffusing capacity of the athletic species. The second effect of the higher hematocrit in the athletic species is even more important: the resulting higher hemoglobin concentration in the capillary blood causes the reaction rate of blood with O_2 to be higher, with the result that the blood component of the diffusing capacity is larger in the athletic species because

Table 7.4 Pulmonary diffusing capacity and aerobic capacity

	$\dot{V}_{O_2}max/M_b$ ml·sec^{-1}·kg^{-1}	$D_{L_{O_2}}/M_b$ ml·sec^{-1}·mmHg^{-1}·kg^{-1}	$[D_{L_{O_2}}]/\{\dot{V}_{O_2}max\}$ mmHg^{-1}
Dog	2.29	0.118	0.052
Goat	0.95	0.080	0.084
D/G	2.4*	1.48*	0.61*
Pony	1.48	0.079	0.053
Calf	0.61	0.050	0.082
P/C	2.4*	1.57*	0.65*
Horse	2.23	0.108	0.048
Steer	0.85	0.054	0.064
H/S	2.6*	2.0*	0.76*
Ath./Sed.	2.5*	1.7*	0.67*

Source: Data from Weibel, Taylor, and Hoppeler (1991).
*Significant ratio.

of both a larger capillary volume and a higher reaction rate. As a result of all this the total pulmonary diffusing capacity $D_{L_{O_2}}$ is about 1.7 times larger in the athletic than in the sedentary species, but this is clearly not matched to the 2.5 times greater $\dot{V}_{O_2}max$.

If we then look at allometric variation, we see that $D_{L_{O_2}}$ and $\dot{V}_{O_2}max$ have different slopes (Figure 7.8): $D_{L_{O_2}}$ increases more steeply with body mass than $\dot{V}_{O_2}max$. As a result, a 30 g animal, like a mouse, has a diffusing capacity per unit body mass that is about 3 times that of a cow of 500 kg, but it must accommodate an O_2 flow rate that is eight times greater. The first conclusion from this is that in allometric variation the diffusing capacity is only partially—about halfway—adjusted to the eightfold difference in $\dot{V}_{O_2}max$, a result similar to that observed between athletic and sedentary species.

Remember that we observed a very similar partial adjustment of the key morphometric parameters at the level of the heart and the microcirculation. There we could resolve the issue by finding other components of the equation that contributed in a significant way to establishing a structure-function match. But because we have used the compound parameter $D_{L_{O_2}}$, which comprises all the major morphometric variables

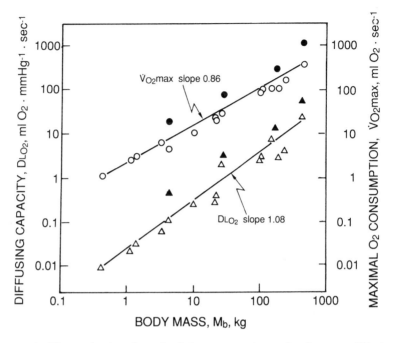

Figure 7.8 Allometric plot of maximal O_2 consumption and pulmonary diffusing capacity against body mass in mammals. Symbols and regressions as in Figure 7.5.

that characterize the gas exchanger, there is no hope of finding other components in this case. The most important conclusion here is that the ratio of the compound design parameter D_{LO_2} to the functional capacity of the gas exchanger $\dot{V}_{O_2}$max is not invariant as our hypothesis had predicted, either in allometric or in adaptive variation. Indeed, the ratio $[D_{LO_2}]/\{\dot{V}_{O_2}max\}$ is about 0.05 mmHg^{-1} in the three athletic species and 0.08 in the sedentary species. This means that the driving force for O_2 diffusion, the difference between the mean partial pressures in alveolar air and capillary blood, $(P_{AO_2} - P_{CO_2})$, amounts to about 20 mmHg in the athletic and 12 mmHg in the sedentary species so that the athletic species operate their gas exchanger with a higher driving force. The athletic species achieve this mainly by a lower mixed venous P_{O_2} and a shorter transit time through the lung capillary, as discussed in Chapter 4 (Figure 4.10).

As a result, the predictions derived from the hypothesis of symmorphosis are not supported, because we cannot find an invariant ratio between diffusing capacity and the aerobic capacity of the system that makes sense—except the seemingly trivial case involving all variables noted in Box 7.1, row 2, such that the invariant ratio is

$$[D_{L_{O_2}}]/\{\dot{V}_{O_2}\text{max} \cdot (P_{A_{O_2}} - P_{c_{O_2}})\}.$$

I say "seemingly trivial" because it is indeed uncertain whether the pressure gradient ($P_{A_{O_2}} - P_{c_{O_2}}$) is not a variable that itself depends on design parameters other than those that compose $D_{L_{O_2}}$. One such variable could be related to the design of the acinar airway system that supplies oxygen to the alveolar surface by diffusion from the bolus of fresh air inspired with each breath. In Chapter 6, I noted that the supply of O_2 to the alveolar surface involves diffusion of O_2 through the residual air in the acinus and that the diameter of the acinar airways is kept larger than predicted in order to improve the diffusion conditions (Figures 6.6 and 6.7). Because the acinar airways are enwrapped by alveoli with their gas exchange surface (Figures 4.2–4.4), O_2 is being extracted from the acinar air all along the acinar airway system, so that the P_{O_2} in the acinar air falls slightly but gradually from the entrance to the acinus to the most peripheral alveoli. What enters the equation in Box 7.1 (row 1) as $P_{A_{O_2}}$ is the *mean* partial pressure of O_2 in all alveoli from the first to the last along the acinar airways. And because they branch by dichotomy over about eight generations (Figure 6.4), the majority of these alveoli are at the periphery and thus the P_{O_2} in the most peripheral airways weighs very heavily. Now, there is good evidence that the total length of the acinar airways is much shorter in smaller species than in larger ones, which may well cause the mean alveolar P_{O_2} to be higher in a mouse than in a cow. This higher pressure head for diffusion may explain why a mouse can maintain a higher oxygen flux rate across its diffusing capacity than a cow. But this is still speculative.

The capillary P_{O_2}, in contrast, depends on the length of the capillary bed and the mean capillary transit time, as outlined in Chapter 4. There I discussed the finding that the higher blood flow in the athletic species causes shorter mean transit times through the lung and accordingly a lower mean $P_{c_{O_2}}$; this because they need the entire capillary path length to load oxygen onto their erythrocytes—unlike the sedentary species, where oxygen loading is completed well before the blood leaves the capillaries (Figure 4.10). Although at present we have no measurements available, it is possible, if not likely, that the capillary path is shorter in small species than in large ones, with the result that their mean Pc_{O_2} would be lower. Thus two design features related to body size may indirectly affect the diffusion conditions in the lung: a shorter acinar path length in small species may cause mean alveolar P_{O_2} to be higher, whereas a shorter capillary path could result in a lower mean capillary P_{O_2}, with the overall result that the driving force for diffusion ($PA_{O_2} - Pc_{O_2}$) would be larger in small species. More work is needed, however, before we can really understand these effects.

Returning to the question raised as part of our test of symmorphosis, we must conclude that the lung's gas exchange structures are adjusted to functional capacity, but apparently incompletely. Therefore the predictions from symmorphosis are not strictly supported in this case. In a way it is difficult to accept that the lung should be "less well" designed than the other parts of the pathway for O_2. After all, we have found the lung to be constructed according to such refined design principles as reflected in Murray's law for conducting airways, and in fractal geometry for the entire airway and vascular systems. To what avail, we must ask, if the lung is overall only crudely adjusted to what the body needs? I cannot but suspect that something may be missing in our arguments, that an important issue is eluding us. There is definitely more work in store, possibly along new lines of inquiry.

Does Symmorphosis Prevail in the Respiratory System?

We have now completed our journey along the pathway for oxygen, wandering upstream from the mitochondria to the lung. We have used this journey to test the hypothesis of symmorphosis, which allowed predictions about what should be considered economic or optimal design to support the capacity for oxidative metabolism in the muscle. What can we conclude?

1. As worked out in the preceding chapters, we consistently found that structural design parameters determine functional capacity at all levels of the system.

2. Structural design parameters are adjusted to differences in functional demand, in adaptive as well as in allometric variation.
3. This adjustment may not be simple, however, as the burden of adjustment can be shared by different structures affecting the same functional step—this sharing of effort indeed hints at economy as a main design constraint.
4. The hypothesis of symmorphosis predicts that invariant ratios exist between design and functional capacity; this we found to be supported for all internal compartments of the body, from the mitochondria to the heart.
5. But this hypothesis was not found to be supported unconditionally for the lung, the structure forming the interface to the environment.

We also predicted that there should not be a single bottleneck in the pathway for O_2 if symmorphosis prevailed. If there was a single bottleneck, we predicted that only this step in the pathway would be adjusted to the overall functional capacity in allometric and adaptive variation, with the other steps varying in an arbitrary way. This is not what we found. At all steps we found that the design components of functional capacity were quite closely related to overall functional capacity—and this is to some extent even fulfilled in the lung—thus supporting symmorphosis by refuting the predictions that would have supported a single bottleneck.

The main deviation from this general conclusion is that the lung maintains significant excess capacity in most species, including the human. Is there something special about the lung that could explain this? Two aspects may be important. The first one is the location of the lung at the interface between the inner world of the body and the environment. In that sense the lung must "protect" the interior world from the unpredictabilities of the outer world. One of these variations is the P_{O_2} in ambient air that is variable with altitude. At high altitudes, as experienced by the pronghorn in the Rocky Mountains, the P_{O_2} in ambient and inspired air may be as low as 100 mmHg compared with 160 at sea level. As a consequence the pressure head for O_2 diffusion cannot be as high as at sea level, so that the excess diffusing capacity relative to gas exchange at sea level will allow $\dot{V}_{O_2}$ max to be maintained even if ambient P_{O_2} is low. Some redundancy in $D_{L_{O_2}}$ gives animals an additional range of freedom in choosing where they live. That this line of thought is not unreasonable may be shown by the finding on the mole rat *Spalax ehrenbergi,* which lives and works hard under severely hypoxic conditions in its underground burrows in Israel. This animal has a lung diffusing capacity that is 50% excessive compared with its $\dot{V}_{O_2}$max, but it is essential for survival; it allows the animal to tolerate the low O_2 concentrations that prevail while it is digging underground in tough soil.

A second aspect to be considered is the very limited malleability of lung structure. In the adult it proves almost impossible to increase the alveolar surface area when the needs for such an increase occur. Whereas heart and mitochondria increase their volume as a consequence of exercise training to augment O_2 delivery and consumption, the pulmonary diffusing capacity remains unaffected. To have excess D_L therefore allows adjustments of the internal components of the pathway, at least up to the point where D_L becomes the ultimate limiting factor, not of $\dot{V}_{O_2}$ but of the upregulation of $\dot{V}_{O_2}$max, by making more structure.

Perhaps the preceding discussion can be taken simply to mean that economy in the sense of symmorphosis prevails in the design of structures supporting functions within the bounds of the organism, within its well-controlled inner world. But the body may have to be wasteful to a certain extent when striving for survival in an unpredictably variable outside world.

8: Adding Complexity in Form and Function: The Combined Pathways for Oxygen and Fuels

The pathway for oxygen discussed in Chapter 7 is a fairly simple system which can be modeled by a single chain of linked steps, each with a specific task but all serving the same function. In this system there are no "escape paths." The only, though important, exception occurs at the very end of the pathway, because ATP can also be generated anaerobically. In fact, it was this option for obtaining ATP by a route other than oxidative phosphorylation that allowed us to estimate the level at which oxidative metabolism becomes limited: when the energy output exceeds that provided by $\dot{V}_{O_2}$max, the safety mechanism of anaerobic glycolysis comes into play (see Chapter 2, especially Figure 2.3). Thus when lactate concentration in the blood increases we know that the limit of $\dot{V}_{O_2}$max is reached. A further feature of this simple pathway is that the oxygen flux rates through all the linked steps are the same under steady-state conditions, and it was therefore easy to test the question whether the design features of the structures supporting this pathway were quantitatively adjusted to the overall function they support, as postulated by symmorphosis.

This is, in fact, not typical for other functional systems. The circulation of the blood, for example, serves many other functions besides the transport of oxygen to the muscles. Most important, perhaps, circulation also supplies nutrients and fuels to the working cells, be they in muscle, brain, or glands.

The relation between design and function is therefore, in general, not as simple as we made believe in the example of the pathway for oxygen. Is it then also possible to test the principle of symmorphosis—the quantitative match between structural design and functional demand—with respect to structures that serve multiple functions and with respect to pathways that are not simple linear chains but for which we should rather conceive models in the form of networks?

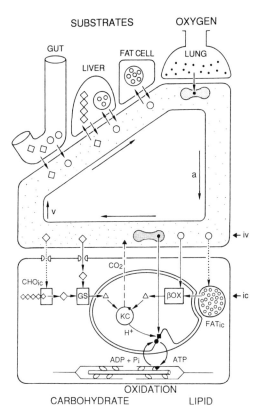

Figure 8.1 Network model of the oxygen and substrate pathways from the supply organs (lung, gut, liver, fat cells) through the circulation (a, arterial; v, venous) to the muscle cells (bottom) with mitochondrion, intracellular fat droplet (FAT_{ic}), glycogen granules (CHO_{ic}), and actomyosin complex which consumes ATP. Substrates are oxidized at the terminal oxidase in the inner mitochondrial membrane (black square), which generates the energy used to phosphorylate ADP to ATP. The oxygen pathway is marked with black dots, the fatty acid pathway with circles, and the glucose pathway with squares. Oxygen is supplied to the blood in the lung; substrates are taken up in the gut; liver and adipose tissue (fat cells) serve as buffers for glucose and fatty acid concentration in the blood. The glucose uptake by the muscle cell is mediated by GLUT transporters (paired hemicircles) in the sarcolemma. The substrate pathways from the microcirculation to the mitochondria are split into a direct and an indirect pathway. The *direct pathways* (full arrows) lead directly to the sites of glycolysis (GS) in the cytosol and of β-oxidation (βOX) in the mitochondria. The *indirect pathways* (dotted arrows) lead through intracellular stores of glycogen and of lipid droplets. The breakdown product of glycolysis and β-oxidation, acetyl-CoA (triangles), enters the Krebs cycle (KC) to generate reducing equivalents (H^+) for oxidation, releasing CO_2 that diffuses to the capillaries (broken arrow) for discharge through the lung. (From Weibel et al. 1996.)

An example of this kind is the combined pathways for the supply of oxygen and fuels to working muscle cells, as outlined in Figure 8.1, a pathway system that uses the circulation of the blood as a common route but is otherwise split to different paths before it converges at the mitochondria. In Chapter 3 we discussed the fact that oxidative phosphorylation in the mitochondria can only proceed if the supply of fuel, specifically glucose and fatty acids, is matched to the rate of oxygen consumption. What we discussed in Chapter 3 at the level of the muscle cells and its associated capillaries has, of course, a necessary link to the entire nutrient supply system. There is an evident parallelism between the oxygen pathway and the fuel pathways in that, in both cases, the substances supplied are taken from the environment: for oxygen through the lung and for nutrients through the gut. Can we now extend the test of the hypothesis of symmorphosis to this more complex system, which has two major input points leading through different pathways that all use the circulation of the blood and converge in the mitochondria where oxygen and fuels are consumed in the process of oxidative phosphorylation? Let us attempt it.

Strategies for Oxygen and Fuel Supply

Although the two pathways serve the same ultimate function, the energetic fueling of muscle work, the strategies for supplying oxygen and substrates for oxidative phosphorylation to muscle mitochondria, are very different. This is, in good part, due to different properties of the substances transferred.

The supply of oxygen is direct from the large oxygen pool in ambient air through the blood to the muscle cells and their mitochondria. Since the body has no capacity for storing oxygen in significant quantities, the uptake of oxygen in the lung and its transport through the circulation are tightly coordinated with the needs of the cells: when the perfusion of muscle increases in exercise due to an increase in cardiac output as well as by redistribution of 90% of the blood flow to the muscles, so is the perfusion of the lung increased, because it always receives 100% of cardiac output. Oxygen uptake and consumption are therefore closely linked both quantitatively and in time.

All this is very different for substrate supply to working muscle.

Substrate uptake and consumption are clearly separate processes that occur in two distinct functional states of the organism at different times. Nutrient uptake in the gut occurs after meals which normally occur intermittently and in animals are followed by a period of rest; but the largest quantities of fuel are burned in the muscle during periods of work, times when animals do not eat. This temporal shift between fuel uptake and consumption occurs for several reasons: food stuffs are not always available to animals in nature, but fuel must be instantaneously available in muscle cells in proportion to the need for energy. Furthermore, food digestion and uptake is not an easy process; it involves several steps of ingestion, digestion, and resorption through the intestinal wall. This requires separate mechanisms for the regulation of fuel uptake and fuel consumption, and this is only possible if intermediate substrate stores are provided.

Design of the Fuel Supply Pathway

For this reason the design of the fuel supply pathway differs in a very significant way from that for oxygen. The latter is made up of an organ of O_2 uptake at the interface to the environment, the lung, where O_2 is transferred by simple diffusion to the blood that then flows directly to the muscle cells where O_2 is delivered, again by simple diffusion. The O_2 pathway can thus be regarded as a two-compartment system—lung and consuming cells—connected by the circulation of the blood. In contrast, the fuel supply pathway is a three-compartment system: fuel storage organs are interposed between the consuming cells and the organs of fuel uptake, the gastrointestinal system (Figure 8.1). The main reason for this basic design is that all mammals have essentially the same requirement for fuels for metabolic energy, chiefly glucose and fatty acids (as outlined in Chapter 3), whereas the food from which these fuels must be derived occurs in nature in very different form. Natural foodstuffs do not offer the simple molecules the cells need. Glucose, for example, is found in plants as disaccharides, two-sugar molecules such as sucrose, but mostly in the form of polysaccharides, long chains of glucose such as starch or cellulose. Even milk sugar is offered to the animals as a disaccharide, lactose, a two-glucose molecule. These molecules cannot be absorbed directly but must first be digested to monosaccharides by special enzymatic processes before the glucose molecules can be taken up in the gut and transferred to the blood. The same is true for fatty acids that can be absorbed but occur mostly in the form of triglycerides, as well as for amino acids that are provided in the form of proteins which cannot be taken up. The intestinal wall, or rather its epithelium, forms a highly selective barrier between the intestinal content, which comes directly from the outside world, and the blood, or the interior milieu of the body.

Design of Nutrient Uptake Systems

To separate the fuel uptake organs from the consumers by storage organs allows a great variety in the design of the gastrointestinal system, and this makes it possible for all food sources in the environment to be utilized. Carnivores eat meat, herbivores eat plants, but a large number of mammals, including humans, are omnivores, capable of feeding on plant as well as animal products. Mammals are all capable of digesting disaccharides and starch, but they are incapable of digesting cellulose, which makes up the large mass of polysaccharides that constitute the plant cell walls, the predominant substance in grass or hay. But many mammals have solved this problem by making use of symbiotic microorganisms that are capable of breaking up cellulose by fermentation; the foremost and also most efficient group here are the ruminants—cows, goats, sheep or antelopes—which harbor these bacteria in their large and complex stomachs whereas other animals, such as the horse or the rabbit, use their large intestine as a fermentation chamber.

Roughly speaking, the gastrointestinal tract of mammals is a tube made of four compartments that are arranged in series: mouth, stomach, small and large intestine. The first compartment is the mouth, where the foodstuffs are ingested and broken down into small fragments by mastication; it is well known that the teeth of a carnivore are different from those of herbivores or omnivores, whose grinding teeth are used to break down the hard plant cell walls to make the cell contents available. The saliva of humans and many animals also contains the enzyme amylase that breaks down starch into the disaccharide maltose, but most of this breakdown occurs later in the gut.

The second compartment, connected to the mouth by the esophagus, is the stomach, a large mixing chamber whose wall is occupied by a large mass of glands that secrete acids and a variety of enzymes to help the breakdown of large foodstuff polymers by acid hydrolysis. This is a relatively slow process; the food is retained in the stomach for some time before it is released in small portions into the gut.

The third and fourth compartments, the gut, form a long tubelike structure of quite variable length—basically made of two subcompartments, the small and large intestine—through which the mixture of partly digested foodstuffs and digestive enzymes is slowly advanced by the action of a muscular sleeve. During this passage carbohydrates, for example, are finally broken down to glucose by a set of special enzymes, amylase and disaccharidases, which are secreted into the gut lumen from the pancreas or from glands in the intestinal wall. In the small intestine the breakdown products are resorbed; in the large intestine much of the water added in the stomach is resorbed back into the blood, and the indigestible parts are "pelleted" into feces.

In our context, nutrient uptake in the small intestine is a particularly significant step. It is effected by the gut epithelium, which is folded up on the villi to enlarge the resorptive surface severalfold over the smooth tube surface of the intestine (Figure 8.2). The membrane of this epithelium is further increased in surface by the formation of microvilli, very regular long cylindrical cytoplasmic formations with a microfilamentous core and a specific membrane (Figure 8.3). This membrane is equipped with transporter molecules, some of which selectively take up the glucose molecules and transfer them to the blood. This glucose uptake is an active process coupled with the exchange of ions across the membrane, and it can occur "uphill," that is, against a concentration gradient, so that glucose can be transferred to the blood at high concentrations. This transfer causes the osmotic pressure at the base of the intestinal epithelium to become transiently very high, resulting in an opening of intercellular junctions; fluid can then pass between the epithelial cells so that additional glucose can be taken up from the gut lumen by what is called solvent drag. The digestion of proteins and of lipids and the uptake of their breakdown products, amino acids and fatty acids, occur in a similar way, by means of specialized enzymes. Whereas amino acids are also taken up at the brush-border membrane and transferred to the blood, the fatty acids enter the vesicular system

Adding Complexity in Form and Function · · · 187

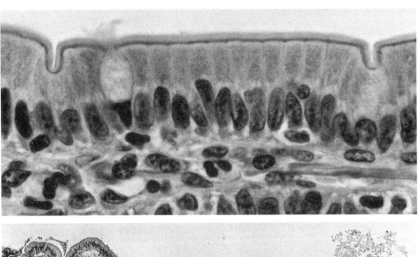

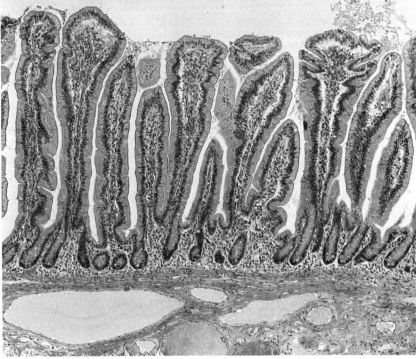

Figure 8.2 Structure of the wall of the small intestine. (a) Intestinal epithelium is made of a simple columnar epithelium of resorptive cells with a brush border interspersed with mucus-producing goblet cells. Note capillaries beneath the epithelium. (b) The surface of the intestinal mucosa, formed by the epithelium, is increased by the formation of villi of varying complexity.

188 · · · Symmorphosis

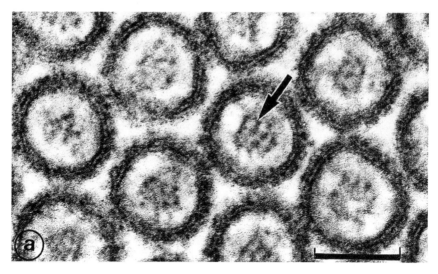

Figure 8.3 Brush border of intestinal epithelial cells is made of long cylindrical microvilli with a filamentous core (arrows) and a surface membrane that carries enzymes and transporter channels for active uptake of glucose, amino acids, and fatty acids. (a) microvilli in cross-section at 200,000x magnification (scale marker: 0.1 μm); (b) in longitudinal section at 36,000x (scale marker: 0.5 μm).

of the epithelial cell and are partly resynthesized to triglycerides to be transferred to the lymphatics rather than to the blood capillaries.

It is most important to note that the blood that leaves the intestinal capillaries enriched in nutrient molecules first passes through the liver (Figure 8.1) by way of the portal vein. This gives the liver—the major metabolic center of the body and at the same time the major storage organ for carbohydrates besides muscle—the opportunity to intercept much of the glucose that is absorbed and to convert it by polymerization into glycogen as the main storage form of carbohydrates (Figure 8.4). This occurs under the control of insulin, the hormone that regulates the blood sugar content and that, when disturbed, can cause diabetes. Not all the glucose is absorbed in the liver, however, so that a large part of the glucose is taken up by muscle cells, again under the control of insulin, and converted into glycogen as intracellular carbohydrate store, as discussed in Chapter 3 (see Figure 3.1).

These processes are modified in ruminants in several important ways that allow them to feed mainly on roughage, cellulose-rich grasses or

Adding Complexity in Form and Function · · · 189

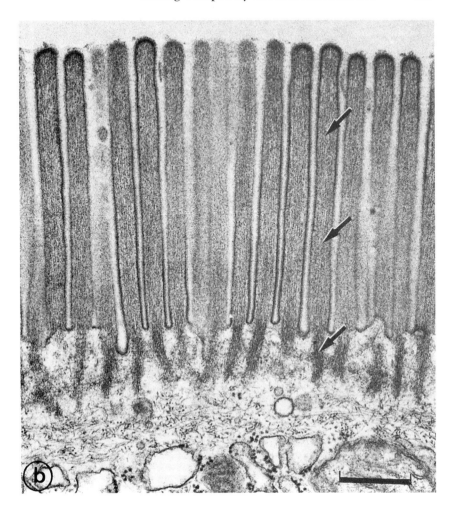

leaves that are, in principle, indigestible for mammals because they do not have enzymes capable of catabolizing these structural plant polysaccharides. In ruminants the stomach is made of several chambers. There a population of symbiotic microorganisms, mostly bacteria and some protozoa, is responsible for the fermentation of cellulose by releasing the enzyme cellulase. To help in the difficult process of digesting the coarse cellulose fibers ruminants ruminate, that is, they regurgitate the contents of their forestomach and chew the cud very extensively before swallowing it again, the process which has given this group of

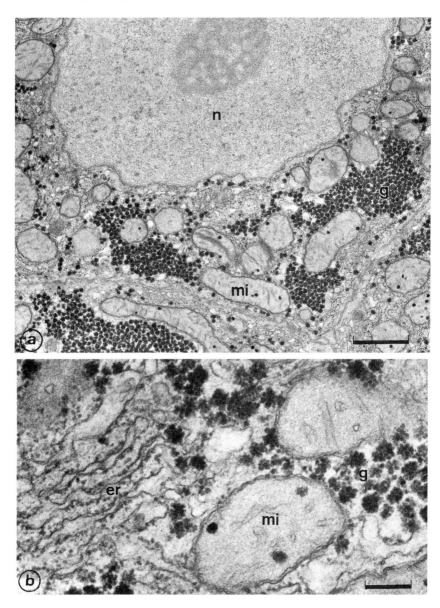

Figure 8.4 Electron micrographs of a liver cell with an agglomeration of glycogen granules (g) near the mitochondria (mi) and endoplasmic reticulum membranes (er) at low (a) and high magnifications (b). Scale markers: (a) 1 μm; (b) 0.2 μm.

animals its name. The reswallowed cud is further fermented and then passed into the subsequent chambers of the stomach. The breakdown product of cellulose is what are called volatile fatty acids, mostly acetate and propionate, as well as CO_2 and methane ("biogas"). The volatile fatty acids are easily diffusible and are therefore taken up into the blood mostly in the stomach. This blood also passes through the liver. The three-carbon acid propionate is 90% taken up by the liver cells and converted into glucose in the process of gluconeogenesis; this glucose is either deposited in the hepatocytes as glycogen or released to the blood to reach other cells, such as muscle, where it is again used to build up the glycogen stores. The two-carbon acetate is mostly converted to fatty acids, and this process occurs to some extent in the liver and to a greater part in fat tissue and in muscle cells, where the fatty acids are deposited as triglyceride droplets. Remember that in the catabolism of fatty acids acetate molecules are clipped off the fatty acid chain to be transferred to acetyl-CoA for entry into the Krebs cycle (see Box 2.1); lipogenesis from acetate is the reverse process.

In other herbivores, such as the horse or the rabbit, the stomach and small intestine are basically the same as in omnivores. They use their large intestine as a fermentation chamber in which microorganisms break down fibrous materials that are left undigested after passage through the stomach and small intestine. These animals have an enlarged colon to serve this function.

The interesting conclusion here is that the metabolic energy made available to the muscle cells, and to other consuming cells as well, is the same irrespective of the nutrient origin, primarily glucose and fatty acids. This is achieved by inserting metabolically active storage organs between the organs of fuel uptake and those of fuel consumption. As a consequence, great variability is allowed in the design of the fuel-uptake organ system with major qualitative differences to cope with the great variety of foodstuffs that the organic world of nature offers. This has allowed the mammals—maybe I should say the whole animal world—to adapt to a very large number of different niches, with specialists feeding on fruits, grasses, or animals and even cadavers, thus keeping the biosphere in balance, leaving essentially no waste. It seems most important in the context of this book to note that these *qualitative* changes occur in the organ system that forms the interface between the inner and the outer world of the body. It is these interface organs that are of paramount importance for the survival of the animal and of the species in the environment, and are therefore subjected to strong selective pressure during evolution. In contrast, the organs and tissues of the inner world appear to be very conservative indeed, and it appears that the most conserved design element is the "innermost" organelle, the mitochondrion, as discussed in Chapters 2 and 3.

Qualitative plasticity, as described above for the gastrointestinal tract, is not the primary topic of this book, which focuses rather on *quantitative* aspects of the malleability of structures to functional needs. Can this also be observed in the gastrointestinal system? If we follow the strategy used in the previous chapters then we would need to study the quantitative structural design in relation to functional conditions that stress the system to the limit. For the gut this condition cannot be achieved by muscular exercise, mainly because of the temporal shift between the uptake of fuels and their combustion in the muscle cells. There are conditions in which fuel consumption at a high rate is continuously required, however, so that the rate of fuel uptake is nearly matched to the rate of fuel consumption. Such conditions include the exposure of the body to very cold environmental temperatures, where thermoregulation requires continuous combustion of fuels, and lactation, where large quantities of sugar, fat and proteins are required not for combustion but for milk production in the mammary glands.

Let us look at this for the case of glucose uptake in the gut epithelium. As mentioned briefly above, this process is mediated by a specific transporter molecule that is inserted in the apical plasma membrane of the intestinal epithelial cell, which forms very regular long microvilli that constitute the brush border of the intestinal epithelium (Figure 8.3). By the formation of these microvilli the epithelial surface becomes increased by up to twentyfold or more. The surface of the intestinal mucosa which is lined by this epithelium is also highly augmented by the formation of villi (Figure 8.2). On the whole, therefore, the inner surface of the gut is very large indeed: in the human it is estimated that the epithelial surface of the small intestine, where most of the nutrient absorption occurs, is about 4 m^2, that is, about twice as large as the skin surface, and this is augmented, due to the formation of microvilli, by a factor of say 20 to yield a membrane surface of some 80 m^2 on which the glucose transporter molecules can be accommodated. Do we need that large a surface? It seems so. It has been shown that the number of transporter molecules per epithelial cell is on the order of 10^6–10^7 and that this number approximately doubles when mice are changed from a no-carbohydrate diet to one rich in carbohydrates, and this is accompanied by a lengthening of the microvilli, that is, by an increase in the membrane surface. This suggests that the epithelial membrane harbors just the amount of transporter molecules it requires and that it is capable of rapid adjustment in their quantity when the diet requires a greater amount of sugar uptake. The capacity for glucose uptake by the gut epithelium therefore appears determined by the epithelial membrane surface together with the density and maximal activity rate of the transporter molecules. This is analogous to the situation in the sarcolemma of muscle cells, as discussed in Chapter 3.

Can this also be tested by varying the fuel or substrate demand of the organism? One such condition is lactation, where the production of milk in the mammary glands is a continuous process that runs in parallel with food digestion and nutrient uptake, at least in animals that have sufficient food available. Indeed, lactation requires such large quantities of nutrients that the gut is put to the test, and in this instance one can therefore explore the question whether the intestine is designed according to symmorphosis. Kimberly Hammond investigated this in a very elegant experiment, in which she attempted to maximize milk production by giving lactating mice up to 14 pups to nurse. She then found that intestinal nutrient uptake was maximized, increased in parallel to the milk output in the mammary glands. Combining the measurement of glucose uptake with estimates of the glucose uptake capacity of the intestinal mucosa, these studies have shown that the intestine normally operates with an excess capacity, or a safety factor, of about two when mice nurse their normal number of about six or seven pups. Adding additional pups first uses this excess capacity for nutrient uptake, but then induces, over a period of a few days, growth processes in the intestinal wall which increase the capacity for nutrient uptake, so that in the end the safety factor is restored. This experiment nicely demonstrates that the nutrient uptake capacity of the intestine can be adjusted to needs, but that it maintains some excess capacity to allow for a transient increase of nutrient uptake above the "normal maximum." Another case is the stimulation of metabolism by cold exposure. When mice are switched from an ambient temperature of 22°C to one of 6°C they more than double their glucose consumption rate, exploiting much

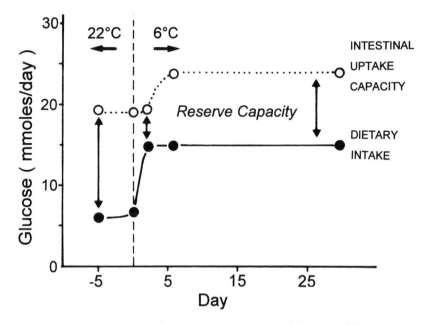

Figure 8.5 Dietary glucose intake in mouse intestine (solid dots and line) compared with the intestinal glucose uptake capacity (open circles, broken line). When ambient temperature is dropped from 22°C to 6°C the dietary intake of glucose immediately increases by more than twofold, exploiting the large reserve uptake capacity (double arrows). After a few days the uptake capacity is enlarged, thus restoring the reserve capacity to some extent. (After Diamond and Hammond 1992.)

of the normal excess capacity of their intestinal transporter system (Figure 8.5); but it takes only a few days until the uptake capacity of the small intestine for glucose is increased by making a larger mucosal surface so that some excess capacity is restored. If lactating mice are additionally exposed to cold the two loads combined cause an even greater increase in glucose uptake capacity. It appears from these experiments that the structural design of the intestinal nutrient uptake system follows the rules we have set for economic design according to symmorphosis. Quite a bit of additional work is required, however, to understand exactly how this system is fine-tuned to the requirements of the body and to the foods that are available—which, in nature, may not be constant throughout the year, for example.

The Substrate Pathways for Fueling Muscle Work

Let us return to the problem of fuel supply to muscle mitochondria commensurate to the substrate needs of oxidative phosphorylation in support of muscle work. The temporal shift between substrate uptake in the gut and utilization in the muscle has a number of consequences which, in fact, may reflect an organizational strategy based on considerations of economy. The foremost one has to do with the distribution of blood flow and its regulation. At rest, total blood flow (cardiac output) in an adult human is about 5–6 liters. Of this about ¼ is used to perfuse the splanchnic circulation, that is, the gut and the liver, and about the same amount perfuses the entire muscle mass (Figure 8.6). Following a rich meal the part flowing to the gut may be somewhat upregulated at the expense of muscle perfusion; as a result we tend to feel tired in this so-called postprandial phase.

In muscular exercise, total blood flow is increased by about a factor of three or four, but we find that now about 90% of the blood flow is directed to the skeletal muscle (Figure 8.6). In contrast, splanchnic blood flow is drastically reduced to only about 1% of total blood flow or just about $1/_5$ of the resting splanchnic blood flow. The immediate consequence of this is easily understood: during exercise the resorption of nutrients from the intestine becomes drastically reduced as well, and fuel can therefore not be supplied directly from the gut to the muscle at the rate required.

As we have seen, during nutrient absorption glucose, and fatty acids as well, are not consumed by the organism but are rather deposited in suitable stores, in the liver and in adipocytes in fat tissue. This process is regulated by the hormone *insulin* which keeps blood sugar concentration at a constant level even during a meal. In exercise, as muscles extract glucose from the plasma in their capillaries, the glucose exchange between hepatocytes and the blood in liver capillaries is reversed, under the effect of the hormone *glucagon,* and glucose is recruited from the glycogen stores in amounts sufficient to keep plasma glucose concentration constant. It appears that the splanchnic blood flow remaining in exercise is sufficient for this, but only because glucose production of the liver can be kept far below the energetic needs of the muscle cells. We noted in Chapter 3 that no more than 10%–20% of

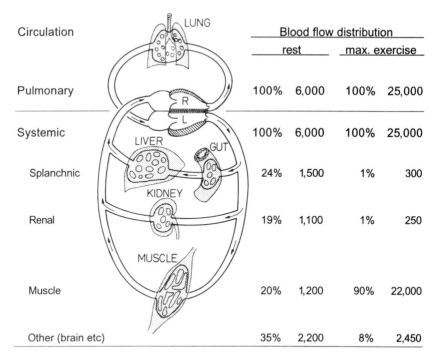

Figure 8.6 Distribution of blood flow to the different tracts of systemic circulation at rest and in maximal exercise. Note that the lung receives 100% of the blood flow in either case. Muscle blood flow is increased 20-fold during exercise whereas blood flow to the intestine and the liver is reduced.

the glucose burnt in the muscle cells during exercise is taken from the blood and thus indirectly from the liver (see Figure 3.4).

The processes are very similar with respect to fatty acids, which become deposited partly in the liver and mostly in adipose tissue. The storage form of fatty acids is as triglycerides, which become aggregated to form spherical lipid droplets of varying size in the cytoplasm of hepatocytes and of adipocytes. During exercise, fatty acids are recruited from these stores by the action of a lipase and discharged to blood plasma, where they are carried bound to albumin. In Chapter 3 we also saw that only about $1/3$ to $1/2$ of the fatty acids burned in muscle cells are taken from the blood and thus indirectly from adipose tissue or the liver (Figure 3.4).

It is thus evident that the supply of substrates to working muscle cells from either the gut or the liver and fat tissue is severely limited; how is it then possible to fuel muscle mitochondria in proportion to the high needs for oxidative phosphorylation in exercise? The solution found by the organism was discussed in Chapter 3: three quarters of the fuels consumed by the mitochondria are drawn from intracellular stores and only one quarter from the blood. We have noted that this may largely be the result of a limited uptake capacity of the muscle cell membrane, the sarcolemma, for both substrates, but mainly for glucose. But this does not pose problems, because glycogen as well as fatty acids are stored in significant quantities in the muscle cells themselves. In fact, the glycogen stores in the total muscle mass are more than twice as large as those in the liver, and they are even several times higher in athletic species, such as the dog. These intracellular fuel stores are stocked up during rest, mainly in the digestive and resorptive period following a meal. This parallels the deposition of glycogen in the liver and of lipid in the fat cells of adipose tissue and is under the control of basically the same hormonal system of insulin. The deposition of substrates in muscle is particularly intense after their glycogen stores have been significantly depleted during a bout of vigorous exercise. Thus the strategy used by marathon runners is to eat large quantities of pasta on the evening before a competition to stock up their muscle cells to the hilt.

The Test of Symmorphosis

These findings have several consequences for our attempt to test the principle of symmorphosis in the combined pathways for oxygen and fuel supply to muscle mitochondria.

1. In contrast to the linear and temporally linked oxygen pathway, the fuel pathway is split both in time and along several parallel axes. The cell has the option of burning either glucose or fatty acids, and it can take these fuels either from the intravascular plasma pool or from intracellular stores (Figure 8.1).

2. The most direct fuel pathway from the gut to the muscle cells is relatively insignificant during exercise. It is operative mostly during rest when, following a meal, the intracellular stores of carbohydrate and lipid are stocked up. An important consequence of this is that in such periods of rest there is plenty of time for transferring the substrates to

the muscle cells; this means, however, that it is difficult to find and measure rate limitations that would allow us to estimate the transfer capacity of this pathway. It is of interest nevertheless that the size of these stores is adjustable; they are larger in well-trained athletes than in sedentary people. We had observed the same difference between the dog and the goat: the dog's muscle glycogen stores are four times as large as those of the goat, and also the intracellular lipid stores are twice as large in the dog as in the goat.

3. During exercise, the important steps occur at the level of muscle tissue, between the capillaries, the intracellular stores and the mitochondria, as discussed in detail in Chapter 3. In this period the organismic stores of glycogen and fatty acids in liver and fat tissue serve to maintain plasma levels, but it turns out that the sarcolemma is such a significant obstacle to fuel supply from vascular sources that most of the fuel is recruited locally from intracellular stores when exercise reaches high levels.

Is it now possible to test the question whether the design of the fuel pathways follows the principle of symmorphosis. Let me give a few arguments why I think this is indeed possible. The test requires that we can measure the functional capacity of the system, which we can then relate to the design parameters of the pathway. The measurement of functional capacity is feasible when "alternate pathways" are available that are switched on when the "primary pathway" has reached its limit or its capacity. Take the example of the oxygen pathway: here the "primary pathway" leads to oxidative phosphorylation at the inner mitochondrial membrane, and the "alternate pathway" is the anaerobic production of ATP through glycolysis (Figure 8.7, top panel). We have estimated the capacity of the oxidative pathway by measuring the point at which energy production switches from oxidative to glycolytic (see Figure 2.3). This measurement allows us to test whether the structures supporting the pathway for oxygen are adjusted to this limit.

In the fuel pathways we found a similar breakpoint as energy needs grow with increasing exercise, occurring when the primary pathway from intravascular sources reaches its limitation and the alternate pathway from intracellular stores is switched on (Figure 8.7, bottom panel). This we can now use to test symmorphosis on the fuel pathways. We shall do this by using the comparative data on exercising dogs and goats presented in Chapter 3.

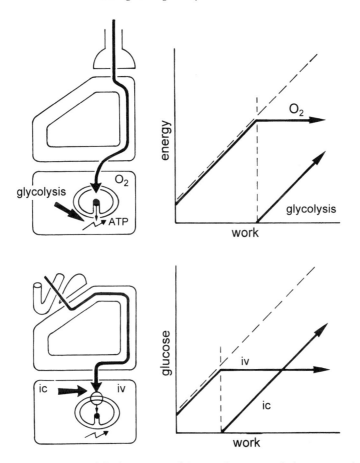

Figure 8.7 Estimation of the limitation of O_2 supply (top) and glucose supply (bottom) to the muscle mitochondria with increasing muscle work. In the pathway for oxygen, limitation occurs when the O_2 flow flattens out and glycolysis provides additional energy, leading to lactate accumulation. In the glucose pathway limitation of vascular supply (iv) is detected when the alternative pathway from intracellular stores (ic) comes into play.

Supply Capacity for Glucose from the Vascular Pool

As an example, let us consider glucose supply to the muscle cells from the pool of circulating glucose, that is, from the "primary pathway" that connects the sources of glucose in liver and gut directly to the muscle cells. I choose this example for the reason that, in Chapter 3, we saw that it is glucose combustion that is increased when exercise intensity increases, whereas fatty acid combustion reaches a plateau at rather low exercise intensities and contributes a small fraction only at high intensities (Figures 3.3 and 3.4).

The first problem is to estimate the fraction of glucose burned that is provided from the plasma pool. This can be done by introducing a known amount of radioactively labeled glucose into the circulation and measuring how much of this is transferred to the muscle cells and oxidized. Such studies showed that the vascular supply of glucose reaches a plateau at 40% maximal exercise intensity in the dog, and at 60% in the goat, beyond which the flux rate of vascular glucose did not further increase (Figure 3.4). It is at these points that the recruitment of glucose from intracellular glycogen stores begins to increase steeply in both species, as shown in Figure 8.8, thus, in fact, confirming the prediction of Figure 8.7. We conclude that the supply path from intravascular glucose sources shows a limited capacity.

Three points are noteworthy:

1. The limitation occurs at two different relative exercise intensities in the dog and the goat. But here it may be interesting to note that the same absolute rate of oxidative phosphorylation per unit muscle mass is reached at 40% maximal exercise intensity in the dog and at 60% in the goat, simply because the dog has a 1.6 times higher capacity for aerobic work. The limitation of vascular glucose supply therefore occurs at the same rate of energy consumption in both species, as measured by O_2 consumption.

2. Furthermore, the vascular glucose supply is limited at about the same rate in the goat and the dog despite a much higher total demand in the athletic species.

3. Perhaps most important, the limitation occurs at a low level of activity and does not limit the achievement of high energy output by the muscles; the additional energy required at higher running speeds is supplied from the alternative pathway, from intracellular stores.

Do these findings help us to answer the question whether the observed limitation of vascular glucose supply is related to design properties of the pathway, as we would predict on the basis of symmorphosis? In order to answer this question we must systematically examine the pathway and attempt to identify potential limiting structures that may form significant barriers to glucose flux. In this it suffices to begin with

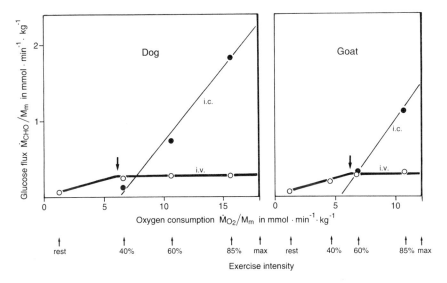

Figure 8.8 Glucose flux from intravascular (iv) and intracellular sources (ic) in dogs and goats. Limitation of vascular supply (arrows) occurs at the same absolute level and the same O_2 consumption in dogs and goats despite a $\dot{V}_{O_2}$max 1.6-fold higher in the dog. (Data from Weber et al. 1996b.)

the microvasculature, because we have already noted that the input of glucose into the blood at higher levels—from the gut and the liver—is adequate to maintain levels of plasma glucose despite its extraction by the working muscles. In the capillaries (Figure 8.9), glucose must traverse the capillary endothelium, and its surface area, $S(c)$, is a potential design parameter that may be limiting the flux rate. Then the interstitium must be traversed, and the average distance that must be covered from the capillary to all points on the sarcolemmal surface is inversely proportional to capillary density: the more capillaries there are on the surface of the muscle cells, the shorter the average distance, which therefore is about inversely proportional to $S(c)$. The last potential barrier is the muscle cell membrane, the sarcolemma, with its surface area $S(sl)$ as a design parameter for potential limitation.

To approach this let us first look at the supply of glucose to muscle at the exercise intensity where the flux appears limited and calculate the flux densities of glucose from intravascular sources, $\dot{M}_{gl}(iv)$, across the unit area of the two chief potential barriers, the endothelium and

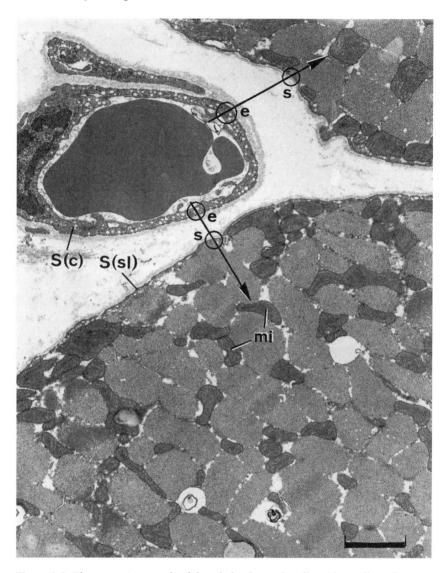

Figure 8.9 Electron micrograph of dog skeletal muscle cells with capillary shows pathway for glucose supply from blood plasma to mitochondrion (mi) with membrane barriers in endothelium (e) and sarcolemma (s) indicated by circles. Morphometric parameters are the capillary surface, S(c), and the sarcolemmal surface S(sl).

the sarcolemma (Table 8.1). Because the limit occurs at the same level of glucose flux (Figure 8.8), we predict that the barrier that limits glucose flux shows the same flux density between the two species, dog and goat, despite the large difference in glucose needs. We find the flux density across the endothelial surface, $\dot{M}_{gl}(iv)/S(c)$, to be 10 μmol · min^{-1} · m^{-2} in the dog, compared with 17 μmol · min^{-1} · m^{-2} in the goat; this does therefore not agree with the prediction, mainly since the goat, with its lower glucose needs, has a higher flux density from the capillary than the dog. The design of muscle capillaries is therefore not adjusted to the needs for fuel supply. We had shown in Chapters 2 and 3 that they are made commensurate to the needs for O_2 supply at high exercise intensities.

The flux density across the sarcolemma, $\dot{M}_{gl}(iv)/S(sl)$, in contrast, turns out to be 2.5 μmol · min^{-1} · m^{-2} in the dog and 2.8 μmol · min^{-1} · m^{-2} in the goat. This suggests that the observed limitation is with high probability related to the capacity of the sarcolemma to let glucose enter the muscle cells rather than to that of the endothelium.

Does this make sense? It appears so, because cell membranes are well known to be potent barriers to glucose transfer that can be overcome only by building special carriers into the membrane, the family of GLUT transporters discussed in Chapter 3, in particular.

So the next question is whether this means that the design of the sarcolemma is adjusted to the functional needs of the muscle cell for glucose. Not really. It only means that flux is limited by a design parameter, the sarcolemmal surface, without pointing to the reason why the sarcolemma has the measured surface. In a broader perspective it also means that the muscle cell can make do with the sarcolemmal transport capacity it has because it has other means, some of which are more efficient, of providing glucose to its mitochondria, namely from its intracellular glycogen stores. This secondary or alternate pathway is then called upon when the cell's demands exceed the supply offered from the vascular sources, as seen in Figure 8.8. The reason why the sarcolemma is limited at the measured rate is, in fact, unrelated to the needs for glucose supply. The sarcolemma serves many other functions, and the dominant function is presumably the control of excitation-contraction coupling through the membrane potential that is mediated by the sarcolemma. There are several good reasons for believing that the sarcolemmal surface might be the same in the dog and the goat, one

Table 8.1 Glucose flux densities across capillary endothelial and sarcolemmal surfaces at 85% $\dot{M}_{O_2}$max

	Dog	Goat	Units
$\dot{M}_{O_2}^{CHO}(mt)/M_m$	17.9	11.8	mmol · min^{-1} · kg^{-1}
$\dot{M}_{gl}(iv)/M_m$	270	330	μmol · min^{-1} · kg^{-1}
$S(c)/M_m$	26.3	17.1	m^2 · kg^{-1}
$\dot{M}_{gl}(iv)/S(c)$	10.3	19.3	μmol · min^{-1} · m^{-2}
$S(sl)/M_m$	109	118	m^2 · kg^{-1}
$\dot{M}_{gl}(iv)/S(sl)$	2.45	2.78	μmol · min^{-1} · m^{-2}

Source: Data from Vock et al. (1996).
Note: $\dot{M}_{O_2}^{CHO}$ is O_2 consumption due to carbohydrate oxidation; M_m is muscle mass.

being that in these animals with very similar body mass the maximal stride frequency is the same, which means that the contraction frequency of the muscle cells is the same during a run.

The Indirect Route for Glucose Supply

With this strict limitation of vascular glucose supply, the major route for glucose provision to the mitochondria in exercising muscle cells is therefore along the indirect pathway via intracellular glycogen stores (Figure 8.1). I argued in Chapter 3 that this broken pathway is physiologically reasonable because, under conditions of high fuel need, glucose can be recruited from stores in the immediate vicinity of the mitochondria and this avoids the problems of flux limitation by resistances such as that offered by the sarcolemma. The pathway from plasma to mitochondria is broken in time: during periods of rest or low activity, glucose is imported into the muscle cell and deposited in the cytoplasm as glycogen granules; during exercise, glucose can then be recruited from these stores when fuel demands are greater than the sarcolemma can provide (Figure 8.8). In a way, this makes sense.

Now remember that after a meal glucose is absorbed from the gut in such a high concentration that cells cannot tolerate it; in a diabetic, severe consequences of glucose overload occur after a carbohydrate-rich meal if the insulin deficiency characterizing this disease cannot be corrected. In healthy people and animals the excess glucose that is discharged into the blood from the gut is sequestered in the liver and in muscle in the form of innocuous glycogen so as to keep the concentration of plasma glucose within very narrow bounds. The hormone insulin upregulates the uptake of glucose into the liver cells and its polymerization to glycogen, and it does the same in muscle. The sarcolemmal glucose transporter system has two types of transporters: GLUT 1 provides a constant baseline permeability of the sarcolemma for glucose, whereas GLUT 4 is regulated in that it is kept in reserve in a small cytoplasmic pool associated with membrane vesicles to become incorporated into the sarcolemma when the need for higher glucose uptake arises in exercise. In Figure 8.8 the low level of glucose influx into the muscle cell at rest is mediated by GLUT 1, whereas the increase in glucose flux with exercise reflects the addition of GLUT 4 to the transporter system, which is known to increase the glucose transport capacity of the membrane by about a factor of 4.

It is interesting that GLUT 4 is also activated in the sarcolemma under the effect of insulin. Thus when, after a meal, insulin is released from the pancreatic islets to downregulate the concentration of plasma glucose, the glucose uptake capacity of the sarcolemma is increased by about a factor of 4; the muscle cells will then avidly take up glucose and begin to stock up their glycogen stores, because, at rest, the mitochondria do not need that much glucose. Make a quick calculation on the basis of the dog data shown in Figure 8.8. At rest, glucose consumption is about 20 μmol $\cdot$ min^{-1} $\cdot$ kg^{-1} muscle mass, and all of the glucose comes from plasma. If glucose uptake is upregulated after a meal by the addition of GLUT 4, we can assume that about the same level of glucose absorption is achieved as in exercise, the plateau level in Figure 8.8, which amounts to 100 μmol $\cdot$ min^{-1} $\cdot$ kg^{-1}; but since the animal is at rest there is a surplus of glucose supply by about 80 μmol $\cdot$ min^{-1} $\cdot$ kg^{-1} that can be used to stock up glycogen stores. In an hour this leads to the deposition of about 5 mmol $\cdot$ kg^{-1} of glucose in glycogen. This corresponds to the quantity of glucose a dog will have to draw from his glycogen stores in 20 minutes when running at 60% maximal exercise intensity. This is not unreasonable. The dog's total muscle glycogen store amounts to about 70 mmol $\cdot$ kg^{-1}; if these stores were totally depleted the dog would take 14 hours to fill them up again, but total depletion is very rare.

It is now worth noting that, because of the time separation between feeding and muscular activity, even the glucose supplied to working muscle cells from plasma is drawn mostly from glycogen stores that have been built up in the liver after a meal. So it is evident that drawing on intramuscular glycogen stores is more efficient than drawing on liver stores. The pathway is much more circuitous in the latter case because the liberated glucose must be transported by the circulation to the target muscle cells and then be taken up by the sarcolemma before it can be shuttled into glycolysis and supplied to the mitochondria. The glycogen stores of the liver act primarily as a buffer for stabilizing plasma glucose concentration and, through this feature, as a basic supplier of glucose to all cells of the body, including the brain. In addition, the metabolic machinery of the liver cells is capable of what is called gluconeogenesis, that is, the synthesis of glucose from smaller compounds such as lactate that may be released into the blood by muscle cells as the end product of anaerobic glycolysis (see Chapter 3) so that glucose can be reconstituted in the liver. In ruminants, the liver also uses propionate, which results from fermentation of cellulose fibers in the forestomach, for gluconeogenesis and thus provides glucose to the plasma as the fuel with which most cells can operate.

Glycogen granules are structural components of the muscle cells. In the sense of symmorphosis we must therefore ask whether these stores are commensurate to the cells' functional needs. This was discussed in Chapter 3, where we saw that the muscle glycogen stores are about 4 times greater in the dog than in the goat. We noted that this allows the dog, a notorious endurance runner, to keep running for twice as long as the goat before exhausting its fuel stores. We thus conclude that the deficit in fuel supply through the direct route is compensated by designing quantitatively adequate intracellular stores as part of the indirect route.

On Symmorphosis in Complex Pathways

The concept of symmorphosis was originally developed in order to interpret structure-function relations in the pathway for O_2, a *simple pathway* for which a *chain* is an equivalent design model. The pathways for substrate supply are not as simple: we have seen that sequential steps from the blood to the mitochondria branch into parallel or alternate pathways. These are *complex pathways* for which a *network* design with multiple inlets is a more appropriate structural equivalent; some structural links, mainly the circulation, are *shared* by different functions, and the functional substrate fluxes can be *partitioned* to take different routes using different structural entities at different times. Shared steps and partitioned functions introduce a considerable degree of complexity into our model for structure-function relations, and the concept of symmorphosis is not testable in a simple linear fashion.

The feature of *shared structures* introduces particular constraints on the functions served. For example, the sarcolemma serves a number of functions, such as excitation-contraction coupling through ion channels, glucose uptake, fatty acid uptake, and O_2 diffusion; the capillaries serve (1) the supply of O_2 from erythrocytes and of substrates from plasma, but also (2) the removal of CO_2, lactate, and waste products as well as the dissipation of heat. In discussing symmorphosis, we will have to determine whether there is a *dominant function* among those that share a step and how this should determine the quantitative design properties. We would expect shared structures to be quantitatively matched only to their primary function.

With respect to the *partitioning* of functional fluxes between branches of a pathway, an important consideration will be how the fluxes through one particular path (or branch) are constrained. The functional requirements of one pathway may impose limitations on independent pathways with which it shares structures. For example, if sarcolemmal surface area is determined by certain contractile properties and the density of GLUT transporters in the membrane is limited, then glucose uptake will be limited independently of substrate need. Another consideration will be whether the different pathways or branches offer particular advantages or disadvantages. For instance, intracellular substrate stores are close to mitochondria and can thus be accessed directly, but they are more or less rapidly depleted if they are not replenished.

The hypothesis of symmorphosis predicts that no more structure is built at each step in a pathway than is required to meet functional demand. To what extent is this prediction fulfilled?

It is interesting that we found the capillaries to be built for O_2 supply whereas their dimension bears no relation to the substrate supply rate to muscle cells. Here the hypothesis of symmorphosis does not seem supported for the substrate pathway, but it does hold for O_2 supply when hematocrit is taken into account.

In contrast, we found that the sarcolemmal surface area bears no relation to maximal O_2 flux rate, but is strictly proportional to the maximal flux rates of glucose (and of fatty acids) from circulatory sources in plasma. From this we could conclude that the hypothesis of symmorphosis is supported because structure (sarcolemmal surface) and function (glucose uptake rate) are tightly related. However, we should not be misled: the match of structure and function at this step does not mean that the design of the sarcolemma is matched to substrate flow, but rather that substrate flow is limited by a sarcolemma designed primarily for other purposes, namely contraction control. It is important to note that the sarcolemma limits substrate supply at rates that are far lower than the total substrate needs of the cell. This is possible only because the cell has an alternate way of obtaining substrates for oxidation at high work rates, namely from intracellular stores.

We find that the rate of substrate supply from intracellular stores at higher work rates is higher in the dog than in the goat. On the basis of the hypothesis of symmorphosis we predict that the size of the glycogen and lipid stores should be higher in the dog than in the goat, and this is the case. However, we find that the sizes of both the glycogen and the lipid stores appear "overadjusted" in the dog, which has a relatively larger intracellular substrate reserve than the goat. Does this refute the hypothesis of symmorphosis? Not necessarily, because the flux *rate* from intracellular stores is not the correct functional parameter to be considered with respect to fuel stores that need a great deal of time to be stocked up; what is more relevant is total fuel use, which is rate times duration of exercise. To provide the dog with a larger pool of fuel reserves is therefore not in contradiction with symmorphosis.

As a general conclusion we find that the pathways for O_2 and for the two substrates are designed to comply with different constraints, and that this is compatible with the principle of symmorphosis applied to network structures. Although individual pathways may have steps in which structure and function are not quantitatively matched, when both the substrate and the O_2 pathways are considered we find that no excess structure is maintained.

The design of these pathways accounts for the very different nature of O_2 and the two substrates, of which substrates can easily be stored and O_2 not. The design of the pathway for nearly instantaneous supply of O_2 from ambient air to the mitochondria is therefore particularly critical. The foodstuffs that nature offers, however, contain the fuels needed by the cells in different forms that must first be broken up by digestion, a difficult and time-consuming process of variable nature, so that the separation in the times of fuel acquisition and fuel usage makes sense.

More generally, note that food is only intermittently available to animals, whereas O_2 is always available and can be absorbed from the atmosphere even during intensive work. The strategies to set up properly designed pathways for O_2 and substrates must therefore be different. I conclude that this study of complex pathways of energy supply to muscle involving both oxygen and fuels has not revealed any trait that would be in contradiction with the principle of symmorphosis.

9: Symmorphosis in Form and Function: Concepts, Facts, and Open Questions

There is no function without form, not even in the smallest molecular array in the cell. Indeed, for proteins to perform their function they must be precisely folded and shaped; much of what today is called "structural biology" is directed toward elucidating the role of form in the function of proteins. But clearly structural biology does not stop there. The study of biological structure is, in fact, much older than the study of molecular forms. We need only think, for example, of the advances made in the understanding of biology—actual changes in the paradigm of the discipline—that came about when the light microscope was invented or, around the middle of this century, the electron microscope. Or when anatomy was complemented by x-rays and, more recently, nuclear magnetic resonance imaging.

Form is of paramount importance for function at all levels of biological organization, from the cell and its organelles to the organ systems and the organism as a whole. Locomotory performance depends on the form of the limbs and the structure of muscle, from its distribution and size to the fine structure of its myofibrils and mitochondria. Form and function are intimately related in any part of the body at all levels of organization, but their ultimate sense derives from how well the parts relate to the whole integrated organism. Indeed, form is so important because it is structural design that makes a whole of the parts, a whole that has its inner logic but is also successful in interacting with the outside world.

Basically, biological form is founded in the genome. The form of a protein is inherent in its primary molecular structure, which is directly determined by its gene. But this is often not enough: protein form also depends on a proper environment, for example on its embedding into a membrane. Membranes, organelles, cells, tissues, organs, and organ systems are hierarchically structured assemblies of functional units.

What counts is their occurrence in the right place and in the right proportions. And this is only partly determined by genes. Order is primarily passed on from cell to cell when cells divide by mitosis: membranes of the new cells derive directly from membranes of the parent cell, and mitochondria derive directly from mitochondria. Rather than dissociating into a "soup" when they multiply, cells pass on their order; mitotic cell division is a highly ordered process. This is even true at the beginning of life: the basic structures of the cells, their membrane system, their cytoskeleton, and their mitochondria, are all derived from those of the oocyte—the basic cell order is therefore "inherited" from the mother, whereas the father contributes only half of the genes through the sperm. When cells grow they add membranes to the preformed membranes. Part of the blueprint of cell design is therefore physically continuous. This is part of the highly conservative nature of the body's internal constitution and of the internal milieu.

In terms of qualitative design the cells and the organisms are indeed very conservative. What interests us here, however, are quantitative differences in the design of cells and organs as they relate to differences in function, therefore as they relate to the diversity of organisms. From that perspective, I have suggested that organismic design is determined by three major principles: adaptation, integration, and economy. *Adaptation* of function to needs implies adequate design and sizing of the structures supporting the function, to make the machine large enough to do the job—something that can be looked at in the sense of engineering applied to the element of interest. *Integration* of function into a whole requires that the design elements are brought into proper relation to one another on the basis of a blueprint for organismic design as it is set up during development guided by the genetic program of the body. But it also requires that the different parts are coadjusted to one another and to the overall needs of the body in as much as they are all interdependent. So integration and adaptation cannot, in fact, be considered separately. *Economy* is perhaps of a different nature, and yet it is intimately related to adaptation and integration. The principle of economy postulates that the adaptation of an element to its specific function is linked to the rule "enough but not too much"—the body should be thrifty but not stingy in designing its parts. And clearly, a well-integrated system avoids unnecessary excesses in any one part. The reasons for postulating the principle of economy of design are severalfold. I have mentioned the problem of space, of accommodating structures of a high degree of complexity all within a finite space in which one notes that there is "standing room only." But there is also the issue of saving on energy—where the body does a much better job than we humans—because in general, energy, that is, the food supply, is scarce and must be acquired by the animal with a great effort, and also at great cost as it must move about to search for food. So the success of a species in the hard competition for survival will be favored if it can be thrifty with the use of its resources.

Since these three principles are so intimately connected we have combined them into one by formulating the broader principle of *symmorphosis*, which simply says that structural design of the organism and its parts must be commensurate to overall functional needs. This will occur if the *adaptation* of single functions is *integrated into a whole* and if this is done *economically*. There is not much radically new about symmorphosis, except that it means that all three basic principles must be obeyed in concert, and perhaps also that it emphasizes the problem of achieving proper proportions within the whole complexity of the organism. It is therefore submitted as a theory for the quantitative relations between form and function, particularly in higher organisms.

Another principle that is closely related to symmorphosis and the three basic principles is *optimization* of design and function, a principle that is, however, highly debated in its biological significance. Adaptation results, in a way, from an optimization process intended to improve performance of the organism or its parts when conditions change for one reason or another. Darwinian evolution by natural selection of the fittest is basically an optimization process related to the success of an organism in its environment. But controversies exist about whether or not optimization is a significant mechanism in the evolution of species. It is therefore important to note that in the context of symmorphosis, optimization has a slightly different touch from the common view of simply looking for the best performance possible of an individual element. What counts is not perfection in the design of each element in the sense of the engineer, but rather the tendency to achieve a well-balanced optimum for the whole on the basis of what is available in the first place. This will inevitably lead to conflicts between different needs and accordingly may end up as the "best compromise" possible. Optimization in this sense is a process of balancing different needs and constraints; the attainment of "perfect optimality" in an individual trait is not essential. What counts in an integrated system such as an animal body is good integration of the parts to make a successful whole, and this is where symmorphosis appears as a most useful theory worth pursuing and testing.

How to Perform a Test of Symmorphosis: Future Prospects

In attempting a test of symmorphosis various approaches are possible. Perhaps the most rewarding would be to take a broad view of evolutionary biology and consider all the changes in form and function that have occurred as animals developed new traits in the course of evolution, when they were forced to cope with different sets of constraints as they moved into new habitats. In this broad scheme animals could not solve their problems within the established blueprint of basic design and this led, often in an unpredictable way, to new inventions, to discontinuities in design that worked better under the new conditions. This has led to an enormous diversity of forms for often similar functions. The supply of oxygen to the working cells, for example, is achieved with several basically different designs. In insects, air-filled tracheoles (Figure 9.1) allow O_2 to diffuse from ambient air into the muscle cells, in fact, right to the surface of the mitochondria. This seems to work well with small animals and this may well be why insects cannot become larger than about 30 grams. Larger animals could only develop when an efficient O_2 transport system evolved: the circulation of the blood as the O_2 carrier. But this depended on the concurrent evolution of a gas exchanger, gills for animals living in water, fishes and young amphibia, and lungs for air-breathing animals. Air breathing is not possible with gills because the surface tension between the air and the humid tissue surface of the gills leads to the collapse of the fine system of lamellae that constitute the gas exchange surface (Figure 9.2a). So it was the evolution of lungs with their internalized fibrous support system (Figure 9.2b), and with their surfactant (see Chapter 5), that allowed animals to breathe air. It is interesting that many air-breathing fishes develop a lunglike organ either from the gills or from the pharyngeal mucosa. And the evolution of a surfactant system was a crucial step in this discontinuous development of new air-breathing

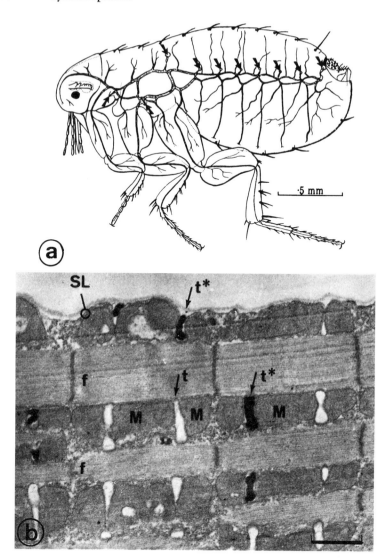

Figure 9.1 The body of an insect is pervaded by a system of tracheae (a). They branch into small tracheoles which penetrate into the flight muscle cells (b) as fine transverse tubules (t) that lie next to the mitochondria (M). In (b) some of the tracheoles (t*) are filled with black oil to show that they communicate with outside air. [From Wigglesworth 1972 (a) and Wigglesworth and Lee 1982 (b).]

species; phospholipid-based surfactants, similar to but different from that found in the mammalian lung, are found in all lung-like organs. This is true even in fish lungs, although these have evolved through different steps than, for example, a mammalian or an amphibian lung. This is a beautiful example of how convergence (the occurrence of the same trait in different places) can point to function.

The broad scope of design variants offered by evolution would appear ideally suited to work out the coadjustment of form and function, but in fact it is quite difficult to do in a systematic way across discontinuities in design such as those mentioned. Discontinuity goes along with major qualitative changes in design that will necessarily require that the models used for the analysis be modified, sometimes even radically changed. Even the bird lung cannot really be analyzed with the same diffusing capacity model as the mammalian lung, because of its fundamentally different design. For these reasons I have concentrated, in this book, on the exploration of mammals, even terrestrial mammals, and this then clearly leaves room for enlarging the scope by studying the systems across major groups of animals in the future. It should indeed be a challenge to look at insects, for example, and test their gas exchange system for symmorphosis. They are a highly successful group of species and cover a size range nearly similar in extent to that of mammals, except that their maximal size lies in the range of small mammals, at about 30 grams. Their muscle cells are basically similar to mammalian muscle, and they produce ATP by oxidation in their mitochondria; so it may be interesting to ask whether the tracheole system does as good a job in delivering O_2 to these organelles as the complex mammalian respiratory system that extends from lung to tissue capillaries. Tracheoles form treelike structures similar to our vessels, but the proportions in the dimensions of the branches seem to be different. So here is a whole program for future research of great biological interest. We can learn much about the functional importance of design by looking for cases where similar design is achieved starting from very different basic structures, what is called convergence. It has been said that convergence in design is a finger pointing to function.

I have taken a different approach, limiting myself to the study of the modulations in form and function as they occur on the basis of an established blueprint. I have focused on mammals, and this has the advantage that changes between the species are largely continuous and

Figure 9.2 Scanning electron micrographs of the gas exchanger: (a) in fish gills with fine lamellae with capillaries standing up, (b) in mammalian lung with alveolar septa with capillaries "hanging" on the fibrous support system.

quantitative rather than qualitative, which allows consistent models of structure-function relations to be tested quantitatively. In mammals there are very few true discontinuities of design, perhaps because the blueprint of mammalian design has been very successful in evolution. The major discontinuity here is in the design of the digestive system, which was necessary to enlarge the range of food availability. The major mass of carbohydrates available in nature is in plant fibers that are indigestible for mammals. These materials have become accessible to some species, ruminants and others, by incorporating in the digestive tract fermentation chambers where symbiotic microorganisms catabolize these fibers; the breakdown products, small volatile fatty acids, can be absorbed in the gastrointestinal system and led into the fuel pool of the organism. The main requirement here is that these products are converted to glucose and fatty acids in the liver so that the other parts of

Symmorphosis in Form and Function · · · 217

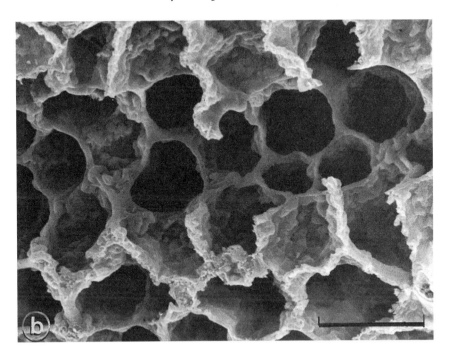

the inner machinery can operate with the same fuels as all mammals. The discontinuity is limited to the mechanism of digestion and the first steps of metabolic conversion in the food-to-fuel chain, whereas the steps from fuel stores to the consumers, in particular the respiratory chain in mitochondria, are identical in all mammals. So this discontinuity does not affect the study of symmorphosis we have followed here, except that we could not enlarge the study to assess the fuel import steps. It clearly would be interesting to extend the studies in this direction by exploring how well the different designs of the gut are serving their functions.

Even limiting myself to mammals, I could not possibly have dealt with all connected parts of the body. I could have either taken the easy approach of looking around for anecdotal evidence, good examples where form and function are closely connected—and there are many— as well as cases where the principle seemingly fails. But that would not have adequately considered the most important principle of integration of function, because the examples would have been disconnected. I have, instead, deliberately chosen the energy supply system as a model case, for several reasons. For one, it is a reasonably well studied system, and I found in C. Richard Taylor a partner particularly well versed in the physiology of energetics. Second, it is a system of reasonable complexity that extends from the cell to the organismic level and furthermore involves several major functional systems such as the locomotor system, the respiratory system, the circulation of blood, and the nutritional system. But even so, the insights we have gained are limited to the systems considered; these findings cannot be indiscriminately extrapolated to other systems. There is still much work to be done before we can claim to have a full picture of the principles governing the relations of form and function in the body.

Conclusions

How well is the principle of *adaptation* of function realized in the energetic system? Let me first note that we have found a number of features which appear to be nonadaptable and thus constitute an internal constraint on adaptation. Perhaps the most interesting nonadaptable trait is the mitochondrial respiratory chain, whose rate of oxidative phosphorylation is invariant among all the species studied, and possibly way beyond these at least to birds, reptiles, and fishes, perhaps even to insects. Maybe this is related to the special nature of this organelle which is considered to be an endosymbiont, deriving from a special bacterium incorporated into the eurkaryotic cell very early in evolution. It is also found that the hemoglobin concentration in erythrocytes is nonadaptable; it has probably achieved its densest possible packing. Among organ functions we found that maximal heart frequency is nonadaptable in the true sense; it varies with absolute body size, and it may well have reached its absolute limit of about 1300 in the smallest mammal, the Etruscan shrew, and in the hummingbird, the smallest bird, both weighing about 2 grams.

We then found that some features have a limited adaptability. For example the concentration of certain enzymes, such as the glucose transporters in plasma membranes or the respiratory chain in the mitochondrial inner membrane, cannot exceed a certain density of packing. If the cell needs to increase its functional performance, this can only be achieved by building more of these membranes. Another example is the concentration of erythrocytes in the blood, which meets an upper limit determined by the flow conditions of the blood through the narrow blood vessels.

In all these cases, overall adaptation is achieved by quantitative adjustment of the structures associated with these features: when more oxidative phosphorylation is needed, mitochondria and their membranes must be augmented; when more substrates must be absorbed in the gut, the membranes of the gut epithelium must be enlarged; and when more oxygen must be transported to the cells, the size of the heart and the density of capillaries on muscle cells must be adjusted. This indicates the importance of introducing a quantitative concept that attempts to assess all possible connected elements in a consistent model, the basis for testing for symmorphosis.

The principle of *integration* plays its role at all levels of organization, and structural design is a most important part of this. The cell is a highly integrated system by design. Most important, we have found the mitochondria to be quantitatively adjusted to the energy needs of the myosin motor in the myofibrils of the muscle fibers, and the intracellular fuel sources essential at high work rates are topographically and quantitatively adjusted to the mitochondrial needs: lipid droplets as the main source for fatty acids are directly apposed to the mitochondria and the glycogen granules are close by, and both are present in adequate quantities. One level higher up, the capillaries with their erythrocytes are tightly adjusted to the muscle cells' O_2 needs, and in this they are adequate for fuel delivery as well, because, in cases of high demand, glucose and fatty acids are drawn in large part from intracellular stores. This example shows that adaptation may use only one, the dominant, function as a yardstick if a structure serves several different functions.

It is evident that form plays a very significant role in the integration of function. A particularly nice example is the way ventilation and perfusion are integrated in the lung by the design of airways and blood vessels. The pulmonary arteries form a tree that precisely follows the airway tree both in course and in dimensions, resulting in two parallel trees whose end twigs converge in the gas exchange region, with the pulmonary veins forming a similar third tree but coursing between the gas exchange units from which they collect the oxygenated blood at the periphery. And then the construction of an extremely thin but vast gas exchange barrier in the lung is the key factor in bringing air and blood together for efficient gas exchange in an integrated, closely matched way.

It is clear that the design of the entire circulation, mainly of the vascular trees in lung and tissues, is directed toward integrating the many varied functions. As we have seen in our examination of the combined pathways for O_2 and fuel supply, the circulation of blood is the common pathway for many functions, many of which we have not discussed. And yet it appears that the size of the heart, the design and dimensions of branching vascular trees, and the packing of red blood cells into blood are primarily gauged to the need for oxygen transport. Why is this? I believe it is because oxygen cannot be stored in significant quantities anywhere in the body or in the cells. This makes O_2 transport from air to cells the most critical process, because O_2 is needed for all cellular processes as the main provider of "clean" energy through oxidative phosphorylation. In these adjustments quantitative modulations of design properties are of central importance because functional parameters, such as heart frequency or enzyme rates, are strictly limited to well-defined maxima. Adaptation of design parameters is therefore the main mechanism for achieving a different function.

Finally, we have attempted the crucial test for integration in the sense of symmorphosis by studying the entire respiratory system and its modifications in instances where the overall limit on performance varied for two reasons: because of body size and because of adaptation for athletic performance. The respiratory system is a simple chain-type model with no escape routes (Figure 9.3). If symmorphosis holds, the O_2 flow through this system should not be limited by a single bottleneck step, but all steps should be coadjusted in design to the altered functional needs. We argued that if there was a bottleneck, only *this* step would be adjusted when the maximal rate of O_2 flow through the system changed. All other steps would either not vary or vary in an arbitrary fashion, because they would have to be designed with abundant excess capacity. This is not what we found. There was no single step that was more precisely adjusted than the others in all the different variations studied. All were coadjusted to the varying needs of the body, including the lung, although its adjustment was not as complete as with

Symmorphosis in Form and Function · · · 221

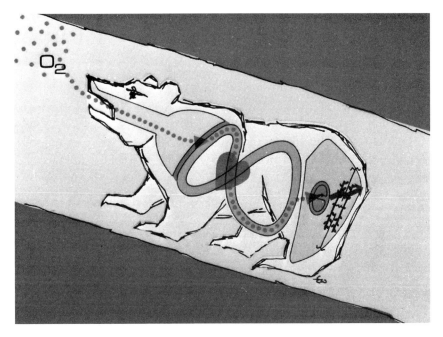

Figure 9.3 The pathway for oxygen is a single path along a chain of organs with no escape route.

the other steps. I have discussed some of the possible explanations in Chapter 7.

We also learned about the importance of coadjustment in the design properties of the different links in the chain of the respiratory system when we studied special cases that depended on adaptation to extreme conditions. One such case was the high-altitude superathlete, the pronghorn, in whom all steps are adjusted, from the lung to the mitochondria, to allow aerobic work at a rate which is twice that of a dog. Another case was the champion of hypoxia tolerance in terrestrial mammals, the blind mole rat from Israel, which has adjusted all parts of its respiratory system so as to be able to perform hard work underground where oxygen is scarce.

The principle of economic design is, in fact, naturally implied in the adaptation of design for efficient function as well as in the integration of functions toward serving the whole: an engine where all parts are coadjusted to one another is also designed economically. Indeed, in the systems studied we have found no instance where waste would have been accepted. We have even found exquisite examples where economy must be an important design principle. For example, in the case of muscle microvasculature the effort of adapting to the higher needs of O_2 supply to the muscle cells in the athletic species is shared between the vessels and the blood. This is achieved by increasing the capillaries only halfway and then increasing the concentration of red cells in the blood to make the other half of the adaptive effort. This case is instructive, because for delivering O_2 to the cells what counts is the number of erythrocytes as O_2 carriers around the muscle cells. Even if the capillaries were increased all the way to match the O_2 delivery needs, the number of erythrocytes would still have to be increased because the larger capillaries contain more blood. So by increasing the hematocrit half the needed capillaries can be saved, at the small price of having the erythrocytes less evenly distributed around the muscle cells. At the same time, to increase the O_2 capacity of the blood has many additional bonuses: cardiac output and particularly the stroke volume of the heart do not have to be adjusted completely in proportion to O_2 needs, and O_2 uptake in the lung is also improved by the higher concentration of red cells on the gas exchange surface. So this is a very economic adjustment. But it would also appear from these arguments that it would be sufficient to simply increase the erythrocyte concentration in the blood and leave the capillaries and the heart unchanged. However, that would introduce serious problems of another kind: the erythrocyte content of the blood would have to be more than doubled, leading to a hematocrit of about 70%; this would make the blood so viscous that very high blood pressure would be required to maintain blood flow. Thus, to strike a compromise between adjusting the vessels and the blood also ensures economic operation of the system.

One of the grand open questions is how symmorphosis comes about. We have here only dealt with the analysis of "final states" of design and have not addressed the issue of "regulated morphogenesis" that is also part of the original definition of symmorphosis. Indeed this is a problem. Regulated morphogenesis occurs to a significant part during fetal development and postnatal growth, at times when the body does not experience the loads for which it must build its systems. The basic design of the lung must be established before birth. Of course, the major quantitative growth occurs after birth, but even then the true maximal load that a system can eventually sustain is never, or very rarely, experienced. Animals do not run voluntarily at $\dot{V}_{O_2}$max—only humans do that. So one of the open questions is how the scope for functional performance is established and how the coadjustment of the different steps we noted is achieved—a vast program, indeed.

Perhaps the most beautiful example of symmorphosis that combines adaptation and integration with economy in a nearly perfect way is the design of the blood vasculature, which determines the distribution of blood and pervades the entire organism. Blood needs to reach every corner of the body, from the skin to the bone and the brain, from the gut to the glands to muscle. Without constant supply of oxygen and fuels, without exchange of other solutes and the removal of wastes, these cells could not persist. Indeed, to maintain the high degree of order that characterizes all units of living organisms, energy must be continuously supplied. So maybe it is no wonder that the circulatory system, mainly the arterial tree, is so perfectly designed.

One of the most fascinating and not yet adequately explored discoveries of recent times is the role that fractal geometry plays in the design of natural structures, as beautifully explained in Benoit Mandelbrot's *Fractal Geometry of Nature*. Fractal geometry is intimately related to the new theories of complexity that are most pertinent in our context. This may indeed be a new perspective in the search for a greater understanding of symmorphosis and the role it plays in elucidating how organisms are made.

We have found that designing the airways in the form of a fractal tree ensures by design that all the millions of gas exchange units in the human lung are at roughly the same distance from the origin. And since the pulmonary arterial tree is designed as a strictly parallel tree until the arteries reach the gas exchange units, ventilation and perfusion become closely matched. It is these features that make it possible to ventilate and perfuse efficiently a surface nearly as large as a tennis court. When we observed that the pulmonary gas exchanger is built with some redundancy, and that some of this excess capacity is used when animals adjust for athletic traits, I suggested that perhaps the lung was not designed as perfectly as the other parts of the respiratory system, that it could be the exception to the rule of symmorphosis. This we cannot exclude, of course, and I have given some possible reasons for it. But, in

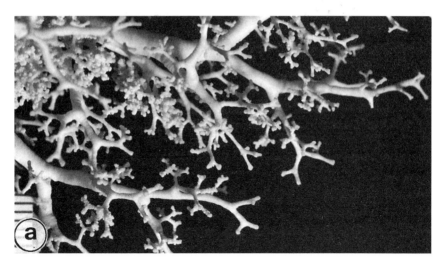

Figure 9.4 Three arrays of branching vessels that abide by Murray's law: (a) the small airways of a human lung, (b) coronary arteries of the human heart, and (c) the astrorhizae imprinted on the inner surface of fossil stromatoporoids, an invertebrate group of uncertain affinities. [(a, b) from Vogel 1998, (c) courtesy M. LaBarbera.]

a way, it is hard to believe that the lung should be suboptimally designed for gas exchange if its internal design shows such examples of perfection as the airway tree and the blood vessels.

With respect to both the lung and the design of the systemic vasculature, the fractal tree model appears to be a natural solution to the difficult problem of how to distribute flows into large surfaces within confined volumes efficiently and in a space-saving way. The principle of fractal tree design is also closely related to the old principles of minimal work for blood flow as worked out by W. R. Hess and S. D. Murray and summarized by D'Arcy W. Thompson. It is a principle employed over and over again in nature, with different variants according to the specific needs. In Figure 9.4 three examples are given where the branching of supply vessels follows Murray's law: the human airway tree, the coronary arteries in the heart wall, and the starlike vessels imprinted in the inner surface of fossil stromatoporoids, a nice example of the way in which biological structures of very different origin show convergent design that can be explained by a tendency to find optimal engineering

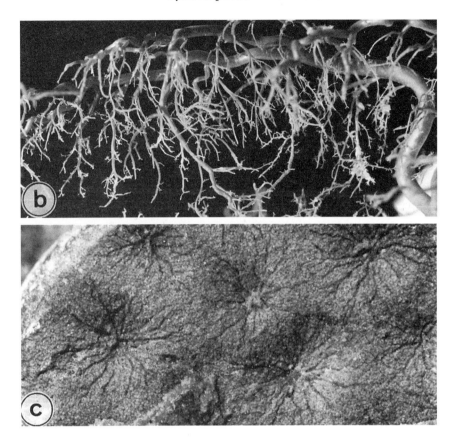

solutions for a critical function. It is then interesting to note that botanical trees likewise show the design of a fractal tree (Figure 9.5), but the size of the branches does not follow Murray's law that the diameter is reduced by the cube root of ½ upon branching. This would indeed not work for getting water from the roots out to the leaves against gravity, because the physical conditions are very different from those reigning in blood vessels, where the blood is pumped under positive pressure through the vasculature.

Maybe, finally, it is permissible to remark that the discovery of fractal structures in the design of the vasculature and the lung may have revealed something about the secret beauty of biological design. Self-similar morphogenesis, which means maintaining the form and the proportions during the growth and proliferation of tissues and organs, is a

226 · · · Symmorphosis

Figure 9.5 The branching pattern of a botanical tree, here a three-year-old sapling of a beech and an end twig of its parent tree, is also fractal, but the proportions between the branches are different from Murray's law. (From Weibel 1994.)

Figure 9.6 The logarithmic spiral, also called "spira mirabilis," describes the growth curve of the nautilus shell (a) and occurs in the complex fractal structures (b) that are found at the periphery of the Mandelbrot set. [(b) from Peitgen and Richter 1986 (map 42).]

Symmorphosis in Form and Function · · · 227

fundamental feature of biological organisms. It finds one of its perfect expressions in the shell of the nautilus (Figure 9.6a), one of the forms that D'Arcy W. Thompson has so admirably described in relation to the laws of growth that determine it. The nautilus, and any ordinary snail as well, show very clear-cut power law growth, which is seen in the shape of the contour curve of the shell: it is a logarithmic spiral in which the angles of the tangent to the curve with the radius are constant, which results in constant proportions in the shape of the chambers. And this design principle allows the perfect solution of an engineering problem: to adjust the floating chamber of the nautilus, the air-filled cells, to the growing size of the animal that sits in the large open-ended compartment. The equiangular or logarithmic spiral, which describes the curve of the nautilus is, by the way, closely related to the golden section—the classical law of proportion where the ratio of the whole to the larger part is the same as that of the larger to the smaller part—which may explain its appealing beauty. The logarithmic spiral is also closely related to the sequence of dimensions in a fractal tree, and it can be plotted very easily into the "Koch tree" shown in Figure 6.8 as a model of the airway tree. And clearly it is very closely related to the beautiful structures that are found at the boundary of the Mandelbrot set (Figure 9.6b), the hallmark of fractal geometry.

It appears that beauty of form and perfection of function are often very closely related in nature.

References and Further Reading

Credits

Index

References and Further Reading

NOTE: To help the reader connect the literature cited below to the text, the references are followed by two-letter abbreviations referring to the chapters' keywords, which are listed at the beginning of each chapter's sources.

1. Form and Function

Keywords: ADaptation ALlometry EConomy EVolution INtegration MOrphometry SYmmorphosis

Further Reading

Alexander, R. McN. 1982. *Options for Animals*. London: Arnold. **AD**

Calder, W. A. 1984. *Size, Function, and Life History*. Cambridge, Mass.: Harvard University Press. **EV, AL**

Garland, T., Jr., and P. A. Carter. 1994. Evolutionary physiology. *Ann. Rev. Physiol.* 56: 579–621. **EV**

Maynard Smith, J. 1978. Optimization theory in evolution. *Ann. Rev. Ecol. Syst.* 9: 31–56. **AD, EV**

Mayr, E. 1991. *One Long Argument: Charles Darwin and the Genesis of Modern Evolutionary Thought*. Cambridge, Mass.: Harvard University Press. **EV**

Rose, M. R., and G. V. Lauder, eds. 1996. *Adaptation*. New York and London: Academic Press. **AD, EV**

Schmidt-Nielsen, K. 1984. *Scaling: Why Is Animal Size So Important?* Cambridge: Cambridge University Press. **EV, AL**

Vogel, S. 1998. Convergence as an analytical tool in evaluating design. In E. R. Weibel, C. R. Taylor, and L. Bolis, eds. 1998. *Principles of Animal Design: The Optimization and Symmorphosis Debate*. Cambridge: Cambridge University Press. Pp. 13–20. **AD, EV**

Weibel, E. R. 1984. *The Pathway for Oxygen*. Cambridge, Mass.: Harvard University Press. **AD, MO, SY**

Weibel, E. R., C. R. Taylor, and L. Bolis, eds. 1998. *Principles of Animal Design: The Optimization and Symmorphosis Debate*. Cambridge: Cambridge University Press. **AD, EC**

Weibel, E. R., C. R. Taylor, and H. Hoppeler. 1991. The concept of symmorphosis: a testable hypothesis of structure-function relationship. *Proc. Nat. Acad. Sci.* 88: 10357–10361. **SY, MO**

References

Alexander, R. McN. 1985. The ideal and the feasible physical constraints on evolution. *Biol. J. Linnean Soc.* 26: 345–358. **EV, AD**

——1998. Symmorphosis and safety factors. In E. R. Weibel, C. R. Taylor, and L. Bolis, eds., *Principles of Animal Design: The Optimization and Symmorphosis Debate*. Cambridge: Cambridge University Press. Pp. 28–35. **AD, SY**

Arnold, S. J. 1992. Constraints on phenotypic evolution. *Am. Naturalist* 140 (Suppl.): S85–S107. **EV**

Cavagna, G. A., N. C. Heglund, and C. R. Taylor. 1977. Mechanical work in terrestrial locomotion: two basic mechanisms for minimizing energy expenditure. *Am. J. Physiol.* 233: R243–R261. **EC, AD**

Currey, J. D. 1984. *The Mechanical Adaptations of Bones*. Princeton: Princeton University Press. **AD**

Darwin, C. 1859. *On the Origin of Species*. London: John Murray. **EV**

Dudley, R., and C. Gans. 1991. A critique of symmorphosis and optimality models in physiology. *Physiol. Zool.* 64: 627–637. **EV, SY**

Dupré, J. 1987. *The Latest on the Best: Essays on Evolution and Optimality*. Cambridge, Mass.: MIT Press. **AD, EV**

Gehring, W. J., M. Affolter, and T. Burglin, T. 1994. Homeodomain proteins. *Ann. Rev. Biochem.* 63: 487–526. **EV**

Gould, S. J., and R. C. Lewontin. 1979. The spandrels of San Marco and the Panglossian paradigm: a critique of the adaptationist programme. *Proc. Roy. Soc. London* B205: 581–598. **EV, AD**

Hess, W. R. 1913. Das Prinzip des kleinsten Kraftverbrauchs im Dienste hämodynamischer Forschung. *Archiv Anat. Physiol.* **EC**

Hofmann, R. R. 1998. How ruminants adapt and optimize their digestive system "blueprint" in response to resource shifts. In E. R. Weibel, C. R. Taylor, and L. Bolis, eds., *Principles of Animal Design: The Optimization and Symmorphosis Debate*. Cambridge: Cambridge University Press. Pp. 220–229. **EV, AD**

Hoyt, D. F., and C. R. Taylor. 1981. Gait and the energetics of locomotion in horses. *Nature* 292: 239–240. **EC**

Parker, G. A., and J. Maynard Smith. 1990. Optimality theory in evolutionary biology. *Nature* 348: 27–33. **EV, AD, EC**

Srere, P. 1998. Molecular symmorphosis, metabolic regulation and metabolons. In E. R. Weibel, C. R. Taylor, and L. Bolis, eds., *Principles of Animal Design: The Optimization and Symmorphosis Debate*. Cambridge: Cambridge University Press. Pp. 131–139. **EV**

Taylor, C. R., and E. R. Weibel. 1981. Design of the mammalian respiratory system. I. Problem and strategy. *Respir. Physiol.* 44: 1–10. **SY, AL, MO**

Thompson, D'A. W. 1942. *On Growth and Form*. Oxford: Oxford University Press. **AD, EC, IN**

Weibel, E. R. 1992. Stereology in perspective: a mature science evolves. *Acta Stereol.* 11: 1–13. **MO**

Weibel, E. R., and C. R. Taylor. 1981. Design of the mammalian respiratory system. *Respir. Physiol.* 44: 1–164. **SY, AD, AL**

Wilson, E. O. 1992. *The Diversity of Life.* Cambridge, Mass.: Belknap Press of Harvard University Press. **EV**

2. Cells and Tissues

Keywords: **CA**pillaries **CO**mparative **EN**ergetics **MI**tochondria **MU**scle **OX**idative phosphorylation **ST**ereology **SU**bstrates **SY**mmorphosis

Further Reading

Balaban, R. S. 1990. Regulation of oxidative phosphorylation in the mammalian cell. *Am. J. Physiol.* 258: C377–C389. **OX**

Hochachka, P. W. 1994. *Muscles as Molecular and Metabolic Machines.* Boca Raton, Fla.: CRC Press. Pp. 1–157. **MU, OX**

Hoppeler, H. 1986. Exercise-induced ultrastructural changes in skeletal muscle. *Int. J. Sports Med.* 7: 187–204. **EN, MI, MU**

———1990. The different relationship of $\dot{V}_{O_2}$max to muscle mitochondria in humans and quadrupedal animals. *Respir. Physiol.* 80: 137–146. **EN, MI**

Hoppeler, H., and S. L. Lindstedt. 1985. Malleability of skeletal muscle in overcoming limitations: structural elements. *J. Exp. Biol.* 115: 355–364. **MU, MI**

Margaria, R. 1976. *Biomechanics and Energetics of Muscular Exercise.* Oxford: Clarendon Press. **EN**

Margulis, L. 1981. *Symbiosis in Cell Evolution.* San Francisco: Freeman. **MI**

McMahon, T. A. 1984. *Muscles, Reflexes, and Locomotion.* Princeton: Princeton University Press. **MU**

Mitchell, P. 1979. Keilin's respiratory chain concept and its chemiosmotic consequences. *Science* 206: 1148–1159. (Nobel Lecture.) **OX, MI**

Somero, G. N., ed. 1998. The biology of oxygen: evolutionary, physiological, and molecular aspects. *J. Exp. Biol.* 201: 1043–1254. **OX**

Tzagaloff, A. 1982. *Mitochondria.* New York: Plenum. **MI**

Weibel, E. R. 1983. Estimating length density and quantifying anisotropy in skeletal muscle capillaries. *J. Microsc.* 131: 131–146. **ST, CA**

———1984. *The Pathway for Oxygen.* Cambridge, Mass.: Harvard University Press. **OX, MI, SY**

References

Bagshaw, C. 1993. *Muscle contraction.* London: Chapman & Hall. **MU**

Conley, K. E., K. A. Christian, H. Hoppeler, and E. R. Weibel. 1995. Heart mitochondrial properties and aerobic capacity are similarly related in a mammal and a reptile. *J. Exp. Biol.* 198: 739–746. **MI, CO**

Conley, K. E., S. R. Kayar, K. Rösler, H. Hoppeler, E. R. Weibel, and C. R. Taylor.

1987. Adaptive variation in the mammalian respiratory system in relation to energetic demand. IV. Capillaries and their relationship to oxidative capacity. *Respir. Physiol.* 69: 47–64.

Crowell, J. W., and E. E. Smith. 1967. Determinant of the optimal hematocrit. *J. Appl. Physiol.* 22: 501–504. **CA**

Cruz-Orive, L. M., and E. R. Weibel. 1990. Recent stereological methods for cell biology: a brief survey. *Am. J. Physiol.* 258: L148–L156. **ST**

Gnaiger, E., B. Lassnig, A. Kuznetsov, G. Rieger, and R. Magreiter. 1998. Mitochondrial oxygen affinity, respiratory flux control and excess capacity of cytochrome c oxidase. *J. Exp. Biol.* 201: 1129–1139. **MI, OX**

Hoppeler, H., H. Howald, K. E. Conley, S. L. Lindstedt, H. Claassen, P. Vock, and E. R. Weibel. 1985. Endurance training in humans: aerobic capacity and structure of skeletal muscle. *J. Appl. Physiol.* 59: 320–327. **EN, MI, MU**

Hoppeler, H., S. R. Kayar, H. Claassen, E. Uhlmann, and R. H. Karas. 1987. Adaptive variation in the mammalian respiratory system in relation to energetic demand. III. Skeletal muscles: setting the demand for oxygen. *Respir. Physiol.* 69: 27–46.

Hoppeler, H., P. Lüthi, H. Claassen, E. R. Weibel, and H. Howald. 1973. The ultrastructure of the normal human skeletal muscle. A morphometric analysis on untrained men, women, and well-trained orienteers. *Pfluegers Arch.* 344: 217–232. **MU, EN**

Hoppeler, H., and E. R. Weibel. 1998. Limits for oxygen and substrate transport in mammals. *J. Exp. Biol.* 201: 1051–1064. **CA**

Howald, H., H. Hoppeler, H. Claassen, O. Mathieu, and R. Straub. 1985. Influence of endurance training on the ultrastructural composition of the different muscle fiber types in humans. *Pfluegers Arch.* 403: 369–376. **MU**

Huxley, A. F. 1980. *Reflections on Muscle.* Liverpool: Liverpool University Press. **MU**

Kayar, S. R., H. Claassen, H. Hoppeler, and E. R. Weibel. 1986. Mitochondrial distribution in relation to changes in muscle metabolism in rat soleus. *Respir. Physiol.* 64: 1–11. **MI, CA, MU**

Kayar, S. R., H. Hoppeler, L. Mermod, and E. R. Weibel. 1988. Mitochondrial size and shape in equine skeletal muscle: a three-dimensional reconstruction study. *Anat. Rec.* 222: 333–339. **MI, MU**

Krogh, A. 1922. *Anatomy and Physiology of Capillaries.* New Haven: Yale University Press. (2nd ed. 1929.) **CA**

Kwast, K. E., P. V. Burke, and R. O. Poyton. 1998. Oxygen sensing and the transcriptional regulation of oxygen responsive genes in yeast. *J. Exp. Biol.* 201: 1177–1195. **OX**

Lindstedt, S. L., J. F. Hokanson, D. J. Wells, S. D. Swain, H. Hoppeler, and V. Navarro. 1991. Running energetics in the pronghorn antelope. *Nature* 353: 748–750. **EN, MU, MI**

Margaria, R., P. Cerretelli, P. E. diPrampero, C. Massari, and G. Torelli. 1963. Kinetics and mechanism of oxygen debt contraction in man. *J. Appl. Physiol.* 18: 371–377. **EN**

Mathieu, O., L. M. Cruz-Orive, H. Hoppeler, and E. R. Weibel. 1983. Estimating length density and quantifying anisotropy in skeletal muscle capillaries. *J. Microsc.* 131: 131–146. **ST, CA**

Mathieu-Costello, O., R. K. Suarez, and P. W. Hochachka. 1992. Capillary-to-fiber geometry and mitochondrial density in hummingbird flight muscle. *Respir. Physiol.* 89: 113–132. **CA, MI, MU**

Mitchell, P. 1961. Coupling of phosphorylation to electron and hydrogen transfer by chemical osmotic type of mechanism. *Nature* (London) 191: 144–148. **MI, OX**

Pette, D., and R. S. Staron. 1993. The molecular diversity of mammalian muscle fibers. *News Physiol. Sci.* 8: 153–157. **MU**

Rome, L. C. 1998. Matching muscle performance to changing demand. In E. R. Weibel, C. R. Taylor, and L. Bolis, eds., *Principles of Animal Design: The Optimization and Symmorphosis Debate.* Cambridge: Cambridge University Press. Pp. 103–113. **MU**

Rome, L. C., R. P. Funke, R. McN. Alexander, G. Lutz, H. Aldridge, F. Scott, and M. A. Freadman. 1988. Why have different muscle fiber types? *Nature* 355: 824–827. **MU**

Rome, L. C., and S. L. Lindstedt. In press. Design of the vertebrate locomotory and muscular system. In W. H. Dantzler, ed., *Handbook of Comparative Physiology.* Oxford: Oxford University Press. **MU**

Saraste, M. A. 1999. Oxydative phosphorylation at the *fin de siècle. Science* 283: 1488–1493. **MI, OX**

Schwerzmann, K., L. M. Cruz-Orive, R. Eggmann, A. Saenger, and E. R. Weibel. 1986. Molecular architecture of the inner membrane of mitochondria from rat liver: a combined biochemical and stereological study. *J. Cell Biol.* 102: 97–103. **MI**

Schwerzmann, K., H. Hoppeler, S. R. Kayar, and E. R. Weibel. 1989. Oxidative capacity of muscle and mitochondria: correlation of physiological, biochemical, and morphometric characteristics. *Proc. Nat. Acad. Sci.* 86(5): 1583–1587. **MI**

Suarez, R. K. 1992. Hummingbird flight: sustaining the highest mass-specific metabolic rates among vertebrates. *Experientia* 48: 565–570. **EN**

———1998. Oxygen and the upper limits to animal design and performance. *J. Exp. Biol.* 201: 1065–1072. **EN, MI**

Suarez, R. K., J. R. B. Lighton, G. S. Brown, and O. Mathieu-Costello. 1991. Mitochondrial respiration in hummingbird flight muscles. *Proc. Nat. Acad. Sci.* 88: 4870–4873. **MI, MU**

Sweeney, H. L. 1998. Fine tuning the molecular motor of muscle. In E. R. Weibel, C. R. Taylor, and L. Bolis, eds., *Principles of Animal Design: The Optimization and Symmorphosis Debate.* Cambridge: Cambridge University Press. Pp. 95–102. **MU**

Taylor, C. R., R. H. Karas, E. R. Weibel, and H. Hoppeler. 1987. Adaptive variation in the mammalian respiratory system in relation to energetic demand. *Respir. Physiol.* 69: 1–127. **SY, EN, MI**

Vock, R., E. R. Weibel, H. Hoppeler, G. Ordway, J.-M. Weber, and C. R. Taylor. 1996. Design of the oxygen and substrate pathways. V. Structural basis of vascular substrate supply to muscle cells. *J. Exp. Biol.* 199(8): 1675–1688. **CA, SU**

Weibel, E. R. 1979. *Stereological Methods.* Vol. I: *Practical Methods for Biological Morphometry.* London, New York, and Toronto: Academic Press. **ST**

Weibel, E. R., C. R. Taylor, and H. Hoppeler. 1991. The concept of symmorphosis: a testable hypothesis of structure-function relationship. *Proc. Nat. Acad. Sci.* 88: 10357–10361.

3. Muscle

Keywords: CApillary FAtty acids GLucose MItochondria MUscle cells OXygen SArcolemmma STores, intracellular SUbstrates SYmmorphosis

Further Reading

Brooks, G. A., and J. Mercier. 1994. Balance of carbohydrate and lipid utilization during exercise: the "crossover" concept. *J. Appl. Physiol.* 76(6): 2253–2261. **EN, GL, FA**

Curry, F. E. 1984. Mechanics and thermodynamics of transcapillary exchange. In E. M. Renkin and C. C. Michel, eds., *Handbook of Physiology.* Section 2, vol. IV: *Microcirculation.* Washington, D.C.: American Physiological Society. Pp. 309–374. **CA**

Hoppeler, H. 1998. The converging pathways for oxygen and substrates in muscle mitochondria. In E. R. Weibel, C. R. Taylor, and L. Bolis, eds., *Principles of Animal Design: The Optimization and Symmorphosis Debate.* Cambridge: Cambridge University Press. Pp. 255–262. **OX, SU, MI, CA, ST, SY**

Margaria, R. 1976. *Biomechanics and Energetics of Muscular Exercise.* Oxford: Clarendon Press. **EN**

Taylor, C. R., E. R. Weibel, J.-M. Weber, R. Vock, H. Hoppeler, T. J. Roberts, and G. Brichon. 1996. Design of the oxygen and substrate pathways. *J. Exp. Biol.* 199: 1643–1709. **OX, SU, MI, CA, ST, SY**

References

Barnard, R. J., and J. F. Youngren. 1992. Regulation of glucose transport in skeletal muscle. *FASEB J.* 6: 3238–3244. **GL, MU**

Crone, C., and D. G. Levitt. 1984. Capillary permeability to small solutes. In E. M. Renkin and C. C. Michel, eds., *Handbook of Physiology.* Sect. 2, vol. IV: *Microcirculation.* Washington, D.C.: American Physiological Society. Pp. 411–466. **CA**

Frøkjaer-Jensen, J. 1985. The continuous capillary: structure and function. *Biologiske Skrifter* 25: 209–253. **CA**

Glatz, J. F. C., G. J. Van der Vusse, and J. H. Veerkamp. 1988. Fatty acid-binding proteins and their physiological significance. *NIPS* 3: 41–43. **FA, SA**

Gollnick, P. D., and B. Saltin. 1988. Fuel for muscular exercise: role of fat. In E. S. Horton and R. L. Terjung, eds., *Exercise, Nutrition, and Energy Metabolism.* New York: Macmillan. Pp. 72–87. **FA, MU**

Goodyear, L. J., M. F. Hirshman, R. J. Smith, and E. S. Horton. 1991. Glucose transporter number, activity, and isoform content in plasma membranes of red and white skeletal muscle. *Am. J. Physiol.* 261: E556–E561. **GL, SA**

Krogh, A., and J. Lindhard. 1920. The relative value of fat and carbohydrate as sources of muscular energy. *Biochem. J.* 14: 290–363. **FA, GL, MU**

Mueckler, M. 1994. Facilitative glucose transporters. *Europ. J. Biochem.* 219: 713–725. **GL**

Pappenheimer, J. R. 1953. Passage of molecules through capillary walls. *Physiol. Rev.* 33: 387–423. **CA**

Perry, M. A. 1980. Capillary filtration and permeability coefficients calculated from measurements of interendothelial cell junctions in rabbit lung and skeletal muscle. *Microvasc. Res.* 19: 142–157. **CA**

Roberts, T. J., J.-M. Weber, H. Hoppeler, E. R. Weibel, and C. R. Taylor. 1996. Design of the oxygen and substrate pathways. II. Defining the upper limits of carbohydrate and fat oxidation. *J. Exp. Biol.* 199(8): 1651–1658. **GL, FA, MU**

Vock, R., H. Hoppeler, H. Claassen, D. X. Y. Wu, R. Billeter, J.-M. Weber, C. R. Taylor, and E. R. Weibel. 1996a. Design of the oxygen and substrate pathways. VI. Structural basis of intracellular substrate supply to mitochondria in muscle cells. *J. Exp. Biol.* 199(8): 1689–1697. **OX, SU, MI, ST, SY**

Vock, R., E. R. Weibel, H. Hoppeler, G. Ordway, J.-M. Weber, and C. R. Taylor. 1996b. Design of the oxygen and substrate pathways. V. Structural basis of vascular substrate supply to muscle cells. *J. Exp. Biol.* 199(8): 1675–1688. **OX, SU, CA, SY**

Weber, J.-M. 1998. Adjusting maximal fuel delivery for differences in demand. In E. R. Weibel, C. R. Taylor, and L. Bolis, eds., *Principles of Animal Design: The Optimization and Symmorphosis Debate.* Cambridge: Cambridge University Press. Pp. 249–254. **SU, MU, SY**

Weber, J.-M., G. Brichon, G. Zwingelstein, G. McClelland, Chr. Saucedo, E. R. Weibel, and C. R. Taylor. 1996a. Design of the oxygen and substrate pathways. IV. Partitioning energy provision from fatty acids. *J. Exp. Biol.* 199(8): 1667–1674. **FA**

Weber, J.-M., T. J. Roberts, R. Vock, E. R. Weibel, and C. R. Taylor. 1996b. Design of the oxygen and substrate pathways. III. Partitioning energy provision from carbohydrates. *J. Exp. Biol.* 199(8): 1659–1666. **GL**

Weibel, E. R., C. R. Taylor, J.-M. Weber, R. Vock, T. J. Roberts, and H. Hoppeler. 1996. Design of the oxygen and substrate pathways. VII. Different structural limits for O_2 and substrate supply to muscle mitochondria. *J. Exp. Biol.* 199(8): 1699–1709. **OX, SU, MI, CA, ST**

4. Organ Design

Keywords: COmparative physiology   DL–diffusing capacity   ENergetics
EVolution   EXperimental   GAs exchanges   MEthods   MOrphology
PHysiology   SYmmorphosis

Further Reading

Dejours, P. 1981. *Principles of Comparative Respiratory Physiology.* 2nd ed. Amsterdam: Elsevier/North-Holland. Pp. 93–108.   **GA, PH, CO**

Dempsey, J. A. 1986. Is the lung built for exercise? *Med. Sci. Sports Exerc.* 18: 143–155.   **EN, PH, DL**

Lindstedt, S. L., J. F. Hokanson, D. J. Wells, S. D. Swain, H. Hoppeler, and V. Navarro. 1991. Running energetics in the pronghorn antelope. *Nature* 353: 748–750.   **EN, CO, PH, MO, SY**

Maina, J. N. 1998. The gas exchangers: structure, function, and evolution of the respiratory processes. In *Zoophysiology,* vol. 37. Berlin and Heidelberg: Springer.   **GA, MO, CO, EV, SY**

Piiper, J. 1994. Search for diffusion limitation in pulmonary gas exchange. In W. W. Wagner, Jr., and E. K. Weir, eds. *The Pulmonary Circulation and Gas Exchange,* vol. 1. Armonk, N.Y.: Futura. Pp. 125–145.   **PH, DL**

Wagner, P. D. 1987. The lungs during exercise. *NIPS* 2: 6–10.   **EN, PH**

Weibel, E. R. 1983. Is the lung built reasonably? *Am. Rev. Respir. Dis.* 128: 752–760.   **DL, MO, CO, SY**

———1984. *The Pathway for Oxygen.* Cambridge, Mass.: Harvard University Press.   **DL, MO, PH, CO, SY**

———1994. Exploring the structural basis for pulmonary gas exchange. In W. W. Wagner, Jr. and E. K. Weir, eds., *The Pulmonary Circulation and Gas Exchange.* Armonk, N.Y.: Futura. Pp. 19–45.   **MO**

———1997. Design and morphometry of the pulmonary gas exchanger. In R. G. Crystal, J. B. West, E. R. Weibel, and P. J. Barnes, eds., *The Lung: Scientific Foundations,* vol. 1. 2nd ed. Philadelphia: Lippincott-Raven Publishers. Pp. 1147–1157.   **DL, MO**

References

Bohr, C. 1909. Ueber die spezifische Tätigkeit der Lungen bei der respiratorischen Gasaufnahme. *Scand. Arch. Physiol.* 22: 221–280.   **DL**

Burri, P. H., and S. Sehovic. 1979. The adaptive response of the rat lung after bilobectomy. *Am. Rev. Respir. Dis.* 119: 769–777.   **DL, EX**

Burri, P. H., and E. R. Weibel. 1971. Morphometric estimation of pulmonary diffusion capacity. II. Effect of Po_2 on the growing lung. Adaptation of the growing rat lung to hypoxia and hyperoxia. *Respir. Physiol.* 11: 247–264.   **DL, EX**

Crapo, J. D., B. E. Barry, P. Gehr, M. Bachofen, and E. R. Weibel. 1982. Cell number and cell characteristics of the normal human lung. *Am. Rev. Respir. Dis.* 126: 332–337. **MO**

Cruz-Orive, L. M. 1979. Estimation of sheet thickness distributions from linear and plane sections. *Biom. J.* 21: 717. **MO, ME**

Federspiel, W. J. 1989. Pulmonary diffusing capacity: implications of two-phase blood flow in capillaries. *Respir. Physiol.* 77: 119–134. **DL, ME**

Fung, Y. C., and S. S. Sobin. 1969. Theory of sheet flow in lung alveoli. *J. Appl. Physiol.* 26: 472–488. **MO, PH**

Geelhaar, A., and E. R. Weibel. 1971. Morphometric estimation of pulmonary diffusion capacity. III. The effect of increased oxygen consumption in Japanese waltzing mice. *Respir. Physiol.* 11:354–366. **DL, EX**

Gehr, P., M. Bachofen, and E. R. Weibel. 1978. The normal human lung: ultrastructure and morphometric estimation of diffusion capacity. *Respir. Physiol.* 32: 121–140. **DL, MO**

Gundersen, H. J. G., T. B. Jensen, and R. Østerby. 1978. Distribution of membrane thickness determined by linear analysis. *J. Microsc.* 113: 27–43. **MO, ME**

Heidelberger, E., and R. B. Reeves. 1990. O_2 transfer kinetics in a whole blood unicellular thin layer. *J. Appl. Physiol.* 68(5): 1854–1864. **GA, ME**

Holland, R. A. B., H. Shibata, P. Scheid, and J. Piiper. 1985. Kinetics of O_2 uptake and release by red cells in stopped-flow apparatus: effects of unstirred layer. *Respir. Physiol.* 59: 71–91. **GA, ME**

Holland, R. A. B., W. van Hezewijk, and J. Zubzanda. 1977. Velocity of oxygen uptake by partly saturated adult and fetal human red cells. *Respir. Physiol.* 29: 303–314. **GA, DL**

Hsia, C. C. W., J. I. Carlin, M. Ramathan, S. S. Cassidy, and R. L. Johnson, Jr. 1991. Estimation of diffusion limitation after pneumonectomy from carbon monoxide diffusing capacity. *Respir. Physiol.* 83: 11–22. **DL, PH, EX**

Hsia, C. C. W., F. Fryder-Doffey, V. Stalder-Navarro, R. L. Johnson, Jr., R. C. Reynolds, and E. R. Weibel. 1993. Structural changes underlying compensatory increase of diffusing capacity after left pneumonectomy in adult dogs. *J. Clin. Invest.* 92: 758–764. **DL, EX**

Hsia, C. C. W., L. F. Herazo, F. Fryder-Doffey, and E. R. Weibel. 1994. Compensatory lung growth occurs in adult dogs after right pneumonectomy. *J. Clin. Invest.* 94: 405–412. **DL, EX**

Hsia, C. C. W., and R. L. Johnson, Jr. 1997. Physiology and morphology of postpneumonectomy compensation. R. G. Crystal, J. B. West, E. R. Weibel, and P. J. Barnes, eds., *The Lung: Scientific Foundations*, vol. 1. 2nd ed. Philadelphia: Lippincott-Raven Publishers. Pp. 1047–1059. **DL, PH, MO, EX**

König, M. F., J. M. Lucocq, and E. R. Weibel. 1993. Demonstration of pulmonary vascular perfusion by electron and light microscopy. *J. Appl. Physiol.* 75: 1877–1883. **MO, ME**

Lindstedt, S. L., J. F. Hokanson, D. J. Wells, S. D. Swain, H. Hoppeler, and V. Navarro. 1991. Running energetics in the pronghorn antelope. *Nature* 353: 748–750. **EN, CO, PH, MO, SY**

Nevo, E. 1991. Evolutionary theory and processes of active speciation and adaptive radiation in subterranean mole rats, *Spalax ehrenbergi* superspecies in Israel. *Evol. Biol.* 25: 1–125. **CO, EV**

Roughton, F. J. W., and R. E. Forster. 1957. Relative importance of diffusion and chemical reaction rates in determining rate of exchange of gases in the human lung, with special reference to true diffusing capacity of pulmonary membrane and volume of blood in the lung capillaries. *J. Appl. Physiol.* 11: 290–302. **GA, DL, PH**

Weibel, E. R. 1963. *Morphometry of the Human Lung*. Berlin, N.Y.: Springer Verlag and Academic Press. **MO, ME**

——— 1970–71. Morphometric estimation of pulmonary diffusion capacity. I. Model and method. *Resp. Physiol.* 11: 54–75. **DL, ME**

——— 1979. *Stereological Methods*. Vol. I: *Practical Methods for Biological Morphometry*. London, New York, Toronto: Academic Press. **ME**

——— 1984. Morphometric and stereological methods in respiratory physiology, including fixation techniques. In A. B. Otis, ed., *Techniques in the Life Sciences. Techniques in Respiratory Physiology*, part 1. County Clare: Elsevier Scient. Publ. Ireland. Pp. 1–35. **ME**

Weibel, E. R., W. J. Federspiel, F. Fryder-Doffey, C. C. W. Hsia, M. König, V. Stalder-Navarro, and R. Vock. 1993. Morphometric model for pulmonary diffusing capacity. I. Membrane diffusing capacity. *Respir. Physiol.* 93: 125–149. **DL, ME**

Weibel, E. R., and B. W. Knight. 1964. A morphometric study on the thickness of the pulmonary air-blood barrier. *J. Cell Biol.* 21: 367–384. **DL, MO, ME**

Weibel, E. R., Marques, L. B., M. Constantinopol, F. Doffey, P. Gehr, and C. R. Taylor. 1987. Adaptive variation in the mammalian respiratory system in relation to energetic demand. VI. The pulmonary gas exchanger. *Respir. Physiol.* 69: 81–100.

Weibel, E. R., C. R. Taylor, J. J. O'Neil, D. E. Leith, P. Gehr, H. Hoppeler, V. Langman, and R. V. Baudinette. 1983. Maximal oxygen consumption and pulmonary diffusing capacity: a direct comparison of physiologic and morphometric measurements in canids. *Respir. Physiol.* 54: 173–188. **DL, CO**

Widmer, H. R., H. Hoppeler, E. Nevo, C. R. Taylor, and E. R. Weibel. 1997. Working underground: respiratory adaptations in the blind mole rat. *Proc. Nat. Acad. Sci.* 94: 2062–2067. **DL, CO, EV**

Yamaguchi, K., D. Nguyen-Phu, P. Scheid, and J. Piiper. 1985. Kinetics of O_2 uptake and release by human erythrocytes studied by a stopped-flow technique. *J. Appl. Physiol.* 58: 1215–1224. **DL, ME, PH**

5. Problems with Lung Design

Keywords: DL–diffusing capacity EXperimental FIber system FLuid (edema) MEchanics MOrphology PAthology PHysiology SF–surface tension (force) SUrfactant

Further Reading

Bachofen, H., and E. R. Weibel. 1996. Pulmonary edema. In P. S. Hasleton, ed., *Spencer's Pathology of the Lung*. 5th ed. New York: McGraw-Hill. Pp. 707–722. **FL**

———1998. Structure-function relationships in the pathogenesis of pulmonary edema. In E. K. Weir and J. T. Reeves, eds., *Pulmonary Edema*. Armonk, N.Y.: Future Publishing. Pp. 17–33. **FL**

Clements, J. A., R. F. Hustead, R. P. Johnson, and I. Gribetz. 1961. Pulmonary surface tension and alveolar stability. *J. Appl. Physiol.* 16: 444–450. **SU, SF**

Mead, J. 1961. Mechanical properties of lungs. *Physiol. Rev.* 41: 281–330. **ME, FI**

Pattle, R. E. 1965. Surface lining of lung alveoli. *Physiol. Rev.* 45: 48–79. **SU, SF**

Weibel, E. R. 1980. Design and structure of the human lung. In A. P. Fishman, ed., *Pulmonary Diseases and Disorders*. New York: McGraw-Hill. **MO**

———1984. *The Pathway for Oxygen*. Cambridge, Mass.: Harvard University Press. Chaps. 9 and 11. **FI, SF, DL**

Weibel, E. R., and H. Bachofen. 1987. How to stabilize the pulmonary alveoli: surfactant or fibers? *News Physiol. Sci.* 2: 72–75. **FI, SU, SF**

———1997. The fiber scaffold of lung parenchyma. In R. G. Crystal, J. B. West, E. R. Weibel, and P. J. Barnes, eds., *The Lung: Scientific Foundations*, vol. 1. 2nd ed. Philadelphia: Lippincott-Raven Publishers. Pp. 1139–1146. **FI, MO**

Weibel, E. R., and J. Gil. 1977. Structure-function relationships at the alveolar level. In J. B. West, ed., *Bioengineering Aspects of the Lung*. New York: Dekker. Pp. 1–81. **FI, SU, SF**

References

Avery, M. E., and J. Mead. 1959. Surface properties in relation to atelectasis and hyaline membrane disease. *Am. J. Dis. of Children* 97: 517–523. **SU, SF, PA**

Bachofen, H., P. Gehr, and E. R. Weibel. 1979. Alterations of mechanical properties and morphology in excised rabbit lungs rinsed with a detergent. *J. Appl. Physiol.* 47: 1002–1010. **SF, EX**

Bachofen, H., S. Schürch, M. Urbinelli, and E. R. Weibel. 1987. Relations among alveolar surface tension, surface area, volume, and recoil pressure. *J. Appl. Physiol.* 62: 1878–1887. **ME, SF, MO**

Bachofen, H., D. Wangensteen, and E. R. Weibel. 1982. Surfaces and volumes of alveolar tissue under zone II and zone III conditions. *J. Appl. Physiol.* 53: 879–885. **ME, SU, MO**

Bachofen, H., J. Weber, D. Wangensteen, and E. R. Weibel. 1983. Morphometric estimates of diffusing capacity in lungs fixed under zone II and zone III conditions. *Respir. Physiol.* 52: 41–52. **DL, EX, SF**

Bachofen, M., and E. R. Weibel. 1977. Alterations of the gas exchange apparatus in adult respiratory insufficiency associated with septicemia. *Am. Rev. Respir. Dis.* 116: 589–615. **FL, PA**

Gil, J., H. Bachofen, P. Gehr, and E. R. Weibel. 1979. Alveolar volume-surface area relation in air- and saline-filled lungs fixed by vascular perfusion. *J. Appl. Physiol.* 47: 990–1001. **MO, EX**

Gil, J., and E. R. Weibel. 1969–70. Improvements in demonstration of lining layer of lung alveoli by electron microscopy. *Respir. Physiol.* 8: 13–36. **SU, MO**

——— 1972. Morphological study of pressure-volume hysteresis in rat lungs fixed by vascular perfusion. *Respir. Physiol.* 15: 190–213. **SF, MO, ME**

Hawgood, S. 1991. Surfactant: composition, structure and metabolism. In R. G. Crystal, J. B. West, P. J. Barnes, N. S. Cherniack, and E. R. Weibel, eds., *The Lung: Scientific Foundations,* vol. 1. New York: Raven Press. Pp. 247–261. **SU**

Neergaard, K. 1929. Neue Auffassungen über einen Grundbegriff der Atemmechanik. Die Retraktionskraft der Lunge, abhängig von der Oberflächenspannung in den Alveolen. *Zeitschr. Ges. Exp. Med.* 66: 373–394. **SF, PH**

Pattle, R. E. 1955. Properties, function, and origin of the alveolar lining layer. *Nature* 175: 1125–1126. **SU**

Schürch, S., J. Goerke, and J. A. Clements. 1978. Direct determination of volume and time dependence of alveolar surface tension in excised lungs. *Proc. Nat. Acad. Sci.* 75: 3417–3421. **SU, ME**

Velazquez, M., E. R. Weibel, Ch. Kuhn III, and D. P. Schuster. 1991. PET evaluation of pulmonary vascular permeability: a structure-function correlation. *J. Appl. Physiol.* 70(5): 2206–2216. **FL, PA, EX**

Weibel, E. R., and J. Gil. 1968. Electron microscopic demonstration of an extracellular duplex lining layer of alveoli. *Respir. Physiol.* 4: 42–57. **SU, MO**

Weibel, E. R., G. S. Kistler, and G. Töndury. 1966. A stereologic electron microscope study of "tubular myelin figures" in alveolar fluids of rat lungs. *Zeitschr. Zellforsch.* 69: 418–427. **SU, MO**

Wilson, T. A. 1979. Parenchymal mechanics at the alveolar level. *Fed. Proc.* 38: 7–10. **ME, FI, SF**

Wilson, T. A., and H. Bachofen. 1982. A model for mechanical structure of the alveolar duct. *J. Appl. Physiol.* 52: 1064–1070. **ME, FI, SF**

6. Airways and Blood Vessels

Keywords: ACinus BLood flow BRonchi BV–blood vessels FRactals LUng structure MOrphogenesis PHysiology VEntilation

Further Reading

Adamson, I. Y. R. 1997. Development of lung structure. In R. G. Crystal, J. B. West, E. R. Weibel, and P. J. Barnes, eds., *The Lung: Scientific Foundations.* 2nd ed. Philadelphia: Lippincott-Raven Publishers. Pp. 993–1001. **MO, LU**

Burri, P. H. 1997. Postnatal development and growth. In R. G. Crystal, J. B. West, E. R. Weibel, P. J. Barnes, eds., *The Lung: Scientific Foundations.* 2nd ed. Philadelphia: Lippincott-Raven Publishers. Pp. 1013–1026. **MO, LU**

Burri, P. H., and E. R. Weibel. 1977. Ultrastructure and morphometry of the developing lung. In W. A. Hodson, ed., *Development of the Lung*. New York and Basel: Dekker. Pp. 215–268. **LU, MO**

Häfeli-Bleuer, B., and E. R. Weibel. 1988. Morphometry of the pulmonary acinus. *Anat. Rec.* 220: 401–414. **LU, AC**

Horsfield, K. 1997. Pulmonary airways and blood vessels considered as confluent trees. In R. G. Crystal, J. B. West, E. R. Weibel, P. J. Barnes, eds., *The Lung: Scientific Foundations*. 2nd ed. Philadelphia: Lippincott-Raven Publishers. Pp. 1073–1079. **BR, BV, BL**

LaBarbera, M. 1990. Principles of design of fluid transport systems in zoology. *Science* 249: 992. **BV**

Losa, G. A., D. Merlini, T. F. Nonnenmacher, and E. R. Weibel. 1998. *Fractals in Biology and Medicine*, vol. 2. Basel: Birkhäuser. **FR**

Mandelbrot, B. B. 1983. *The Fractal Geometry of Nature*. 2nd ed. San Francisco: Freeman. **FR**

Nonnenmacher, T. F., G. A. Losa, and E. R. Weibel. 1994. *Fractals in Biology and Medicine*. Basel: Birkhäuser. **FR**

Weibel, E. R. 1963. *Morphometry of the Human Lung*. Berlin and New York: Springer Verlag and Academic Press. **LU, BR**

———1984. *The Pathway for Oxygen*. Cambridge, Mass.: Harvard University Press. **LU, MO, BR**

———1991. Fractal geometry: a design principle for living organisms. *Am. J. Physiol.* 261: L361–L369. **FR, BR, BV**

———1997. Design of airways and blood vessels considered as branching trees. In R. G. Crystal, J. B. West, E. R. Weibel, and P. J. Barnes, eds., *The Lung: Scientific Foundations*, vol. 1. 2nd ed. Philadelphia: Lippincott-Raven Publishers. Pp. 1061–1071. **BR, BV, FR**

References

Alescio, T., and A. Cassini. 1962. Induction in vitro of tracheal buds by pulmonary mesenchyme grafted on tracheal epithelium. *J. Exp. Zool.* 150: 83–94. **MO, BR**

Burri, P. H. 1984. Lung development and histogenesis. In A. P. Fishman and A. B. Fisher, eds., *Handbook of Physiology: Respiration*, vol. 4. Washington, D.C.: American Physiological Society. **LU, MO**

Burri, P. H., J. Dbaly, and E. R. Weibel. 1974. The postnatal growth of the rat lung. I. Morphometry. *Anat. Rec.* 178: 711–730. **LU, MO**

Hess, W. R. 1914. Das Prinzip des kleinsten Kraftverbrauches im Dienste hämodynamischer Forschung. *Arch. Anat. Physiol.* Physiol. Abt. **BV, PH**

Horsfield, K., G. Dart, D. E. Olson, G. F. Filley, and G. Cumming. 1971. Models of human bronchial tree. *J. Appl. Physiol.* 31: 207–217. **BR**

Kitaoka, H., P. H. Burri, and E. R. Weibel. 1996. Development of the human fetal airway tree: Analysis of the numerical density of airway endtips. *Anat. Rec.* 244: 207–213. **MO, FR, BR**

Kurz, H., and K. Sandau. 1997. Modeling of blood vessel development—bifurcation pattern and hemodynamics, optimality and allometry. *Comm. on Theoret. Biol.* 4: 261–292. **BV, MO, FR**

Kurz, H., K. Sandau, and B. Christ. 1997. On the bifurcation of blood vessels—Wilhelm Roux's doctoral thesis (Jena 1878)—a seminal work for biophysical modelling in developmental biology. *Ann. Anat.* 179: 33–36. **BV**

Murray, C. D. 1926a. The physiological principle of minimum work applied to the angle of branching of arteries. *J. Gen. Physiol.* 9: 835–841. **BV, PH, BL**

——— 1926b. The physiological principle of minimum work. I. The vascular system and the cost of blood. *Proc. Nat. Acad. Sci.* 12: 207–214. **BV, PH, BL**

Post, M. 1997. Tissue interactions. In R. G. Crystal, J. B. West, E. R. Weibel, and P. J. Barnes, eds., *The Lung: Scientific Foundations.* 2nd ed. Philadelphia: Lippincott-Raven Publishers. Pp. 1003–1011. **MO, BR**

Rodriguez, M., S. Bur, A. Favre, and E. R. Weibel. 1987. Pulmonary acinus: geometry and morphometry of the peripheral airway system in rat and rabbit. *Am. J. Anat.* 180: 143–155. **LU, AC**

Spooner, B. S., and N. K. Wessels. 1970. Mammalian lung development: interactions in primordium formation and bronchial morphogenesis. *J. Exp. Zool.* 175: 445–454. **MO, BR**

Suwa, N., T. Niwa, H. Fukasawa, and Y. Sasaki. 1963. Estimation of intravascular blood pressure gradient by mathematical analysis of arterial casts. *Tohoku J. Exp. Med.* 79: 168. **BV, BL, FR**

Verbanck, S., E. R. Weibel, and M. Paiva. 1993. Simulations of washout experiments in postmortem rat lungs. *J. Appl. Physiol.* 75(1): 441–451. **VE, BR, PH**

Weibel, E. R. 1994. Design of biological organisms and fractal geometry. In T. Nonnenmacher, G. A. Losa, and E. R. Weibel, eds., *Fractals in Biology and Medicine.* Basel: Birkhäuser. Pp. 68–85. **FR, BR, LU**

Weibel, E. R., and D. M. Gomez. 1962. Architecture of the human lung. *Science* 137: 577–585. **BR, LU**

West, B. J., V. Barghava, and A. L. Goldberger. 1986. Beyond the principle of similitude: renormalization in the bronchial tree. *J. Appl. Physiol.* 60: 1089–1097. **BR, FR**

Wilson, T. A. 1967. Design of the bronchial tree. *Nature* 1967/2: 668–669. **BR**

7. The Pathway for Oxygen

Keywords: **AE**robic capacity **CA**pillaries **CO**mparative physiology **EN**ergetics **GA**s exchange, pulmonary **HE**art **MI**tochondria **MO**rphology **MU**scle **OX**ygen pathway **PH**ysiology **SY**mmorphosis

Further Reading

Bicudo, J.E.P.W., ed. 1992. *The Vertebrate Oxygen Transport Cascade: From Atmosphere to Mitochondria.* Boca Raton: CRC Press. **OX, CO, PH**

Dejours, P. 1981. *Principles of Comparative Respiratory Physiology.* 2nd ed. Amsterdam: Elsevier/North-Holland. Pp. 93–108. **OX, PH, CO**

Feder, M., A. F. Bennett, W. Burgren, and R. B. Huey, eds. 1987. *New Directions in Physiological Ecology.* New York: Cambridge University Press. **CO, PH**

Hoppeler, H. 1990. The different relationship of $\dot{V}_{O_2}$max to muscle mitochondria in humans and quadrupedal animals. *Respir. Physiol.* 80: 137–146. **OX, AE, MU, CO**

Hoppeler, H., and S. L. Lindstedt. 1985. Malleability of skeletal muscle in overcoming limitations: structural elements. *J. Exp. Biol.* 115: 355–364. **MU, MI, AE**

Jones, J. H., and S. L. Lindstedt. 1993. Limits to maximal performance. *Ann. Rev. Physiol.* 55: 547–569. **OX, PH, AE, CO**

Kleiber, M. 1961. *The Fire of Life: An Introduction to Animal Energetics.* New York: Wiley. **AE, EN, CO**

McMahon, T. A. 1984. *Muscles, Reflexes, and Locomotion.* Princeton: Princeton University Press. **MU, EN, PH**

Suarez, R. K. 1996. Upper limits to mass-specific metabolic rates. *Ann. Rev. Physiol.* 58: 583–605. **AE, OX, CO**

Taylor, C. R., and E. R. Weibel. 1997. Learning from comparative physiology. In R. G. Crystal, J. B. West, E. R.Weibel, and P. J. Barnes, eds., *The Lung: Scientific Foundations,* vol. 2. 2nd ed. Philadelphia: Lippincott-Raven Publishers. Pp. 2043–2054. **OX, AE, SY**

Taylor, C. R., E. R. Weibel, and L. Bolis. 1985. *Design and Performance of Muscular Systems.* Cambridge: The Company of Biologists Limited. **MU, EN, CO**

Taylor, C. R., E. R. Weibel, H. Hoppeler, and R. H. Karas. 1989. Matching structures and functions in the respiratory system. In S. C. Wood, ed., *Comparative Pulmonary Physiology: Current Concepts.* New York: Dekker. Pp. 27–65. **OX, AE, SY**

Weibel, E. R. 1979. Oxygen demand and the size of respiratory structures in mammals. In S. C. Wood and C. Lenfant, eds., *Evolution of Respiratory Processes.* New York: Dekker. Pp. 289–346. **OX, AE, CO**

———1984. *The Pathway for Oxygen: Structure and Function in the Mammalian Respiratory System.* Cambridge, Mass.: Harvard University Press. **OX**

Weibel, E. R., and C. R. Taylor. 1981. Design of the mammalian respiratory system. *Resp. Physiol.* 44: 1–164. **OX, CO, SY**

Weibel, E. R., C. R. Taylor, and L. Bolis. 1998. *Principles of Animal Design: The Optimization and Symmorphosis Debate.* Cambridge: Cambridge University Press. **CO, SY**

Weibel, E. R., C. R. Taylor, and H. Hoppeler. 1991. The concept of symmorphosis: a testable hypothesis of structure-function relationship. *Proc. Nat. Acad. Sci.* 88: 10357–10361. **SY, OX**

———1992. Variations in function and design: testing symmorphosis in the respiratory system. *Respir. Physiol.* 87: 325–348. **OX, SY**

References

Alexander, R. McN. 1988. *Elastic Mechanisms in Animal Movement*. Cambridge: Cambridge University Press. **MU, EN**

Armstrong, R. B., B. Essen-Gustavsson, H. Hoppeler, J. H. Jones, S. R. Kayar, M. H. Laughlin, A. Lindholm, K. E. Longworth, C. R. Taylor, and E. R. Weibel. 1992. O_2 delivery at V_{O_2}max and oxidative capacity in muscles of Standardbred horses. *J. Appl. Physiol.* 73(6): 2274–2282. **OX, MU, PH**

Conley, K. E., S. R. Kayar, K. Rösler, H. Hoppeler, E. R. Weibel, and C. R. Taylor. 1987. Adaptive variation in the mammalian respiratory system in relation to energetic demand. IV. Capillaries and their relationship to oxidative capacity. *Respir. Physiol.* 69: 47–64. **OX, CA**

Constantinopol, M., J. H. Jones, E. R. Weibel, C. R. Taylor, A. Lindholm, and R. H. Karas. 1989. Oxygen transport during exercise in large mammals: Oxygen uptake by the pulmonary gas exchanger. *J. Appl. Physiol.* 67(2): 871–878. **OX, GA**

Evans, D. L., and R. J. Rose. 1988. Cardiovascular and respiratory responses to submaximal exercise in the thoroughbred horse. *Pfluegers Arch.* 411: 316–321. **OX, PH, EN, AE**

Farley, C. T., J. Glasheen, and T. A. McMahon. 1993. Running springs: speed and animal size. *J. Exp. Biol.* 185: 71–86. **MU, EN**

Geelhaar, A., and E. R. Weibel. 1971. Morphometric estimation of pulmonary diffusion capacity. III. The effect of increased oxygen consumption in Japanese waltzing mice. *Respir. Physiol.* 11: 354–366. **GA, MO**

Gehr, P., D. K. Mwangi, A. Ammann, S. Sehovic, G. M. O. Maloiy, C. R. Taylor, and E. R. Weibel. 1981. Design of the mammalian respiratory system. V. Scaling morphometric pulmonary diffusing capacity to body mass: wild and domestic mammals. *Respir. Physiol.* 44: 61–86. **OX, GA, CO, SY**

Gleeson, T. T., W. J. Mullin, and K. M. Baldwin. 1983. Cardiovascular response to treadmill exercise in rats: effects of training. *J. Appl. Physiol.* 54: 789–793. **HE, PH**

Heglund, N. C., M. A. Fedak, C. R. Taylor, and G. A. Cavagna. 1982. Energetics and mechanics of terrestrial locomotion. IV. Total mechanical energy changes as a function of speed and body size in birds and mammals. *J. Exp. Biol.* 97: 57–66. **EN, PH**

Hoppeler, H., J. H. Jones, S. L. Lindstedt, K. E. Longworth, C. R. Taylor, R. Straub, and A. Lindholm. 1987. Relating maximal oxygen consumption to skeletal muscle mitochondria in horses. In J. R. Gillespie and N. E. Robinson, eds., *Equine physiology II*. Davis, Calif.: ICEEP. Pp. 278–289. **AE, MI**

Hoppeler, H., S. R. Kayar, H. Claassen, E. Uhlmann, and R. H. Karas. 1987. Adaptive variation in the mammalian respiratory system in relation to energetic demand. III. Skeletal muscles: setting the demand for oxygen. *Respir. Physiol.* 69: 27–46. **OX, MU, MI, AE**

Hoppeler, H., S. L. Lindstedt, H. Claassen, C. R. Taylor, O. Mathieu, and E. R. Weibel. 1984a. Scaling mitochondrial volume in heart to body mass. *Respir. Physiol.* 55: 131–137. **MI, HE, CO**

Hoppeler, H., S. L. Lindstedt, E. Uhlmann, A. Niesel, L. M. Cruz-Orive, and E. R. Weibel. 1984b. Oxygen consumption and the composition of skeletal muscle tissue after training and inactivation in the European woodmouse (Apodemus sylvaticus). *J. Comp. Physiol.* B155: 51–61. **AE, MU, CO**

Jones, J. H., E. K. Birks, and J. R. Pascoe. 1992. Factors limiting aerobic performance. In J.E.P.W. Bicudo, ed., *The Vertebrate Oxygen Transport Cascade: From Atmosphere to Mitochondria*. Boca Raton: CRC Press. Pp. 169–178. **AE**

Jones, J. H., K. E. Longworth, A. Lindholm, K. E. Conley, R. H. Karas, S. R. Kayar, and C. R. Taylor. 1989. Oxygen transport during exercise in large mammals. I. Adaptive variation in oxygen demand. *J. Appl. Physiol.* 67(2): 862–870. **OX, AE, CO, PH**

Kayar, S. R., H. Hoppeler, J. H. Jones, K. Longworth, R. B. Armstrong, M. H. Laughlin, S. L. Lindstedt, J.E.P.W. Bicudo, K. Groebe, C. R. Taylor, E. R. Weibel. 1994. Capillary blood transit time in muscles in relation to body size and aerobic capacity. *J. Exp. Biol.* 194: 69–81. **CA, AE**

Kayar, S. R., H. Hoppeler, S. L. Lindstedt, H. Claassen, J. H. Jones, B. Essen-Gustavsson, and C. R. Taylor. 1989. Total muscle mitochondrial volume in relation to aerobic capacity of horses and steers. *Pfluegers Arch.* 413: 343–347. **MI, AE**

Kram, R., and C. R. Taylor. 1990. Energetics of running: a new perspective. *Nature* 346: 265–267. **MU, EN**

Lindstedt, S. L. 1984. Pulmonary transit time and diffusing capacity in mammals. *Am. J. Physiol.* 246: R384–R388. **GA, CO**

Lindstedt, S. L., J. F. Hokanson, D. J. Wells, S. D. Swain, H. Hoppeler, and V. Navarro. 1991. Running energetics in the pronghorn antelope. *Nature* 353: 748–750. **EN, OX, CO**

Lindstedt, S. L., and Jones, J. H. 1987. Symmorphosis: the concept of optimal design. In M. Feder, A. F. Bennett, W. Burgren, R. B. Huey, eds., *New Directions in Physiological Ecology*. Cambridge: Cambridge University Press. Pp. 289–309. **SY**

Longworth, K. E., J. H. Jones, J.E.P.W. Bicudo, C. R. Taylor, and E. R. Weibel. 1989. High rate of O_2 consumption in exercising foxes: large P_{O_2} difference drives diffusion across the lung. *Respir. Physiol.* 77: 263–276. **GA, CO**

Mathieu, O., R. Krauer, H. Hoppeler, P. Gehr, S. L. Lindstedt, R. McN. Alexander, C. R. Taylor, and E. R. Weibel. 1981. Design of the mammalian respiratory system. VII. Scaling mitochondrial volume in skeletal muscle to body mass. *Respir. Physiol.* 44: 113–128. **MI, CO**

Mathieu-Costello, O., R. K. Suarez, and P. W. Hochachka. 1992. Capillary-to-fiber geometry and mitochondrial density in hummingbird flight muscle. *Resp. Physiol.* 89: 113–132. **MI, CA, CO**

Prothero, J. 1979. Heart weight as a function of body weight in mammals. *Growth* 43: 139–150. **HE, CO**

Stahl, W. R. 1967. Scaling of respiratory variables in mammals. *J. Appl. Physiol.* 22(3): 453–460. **CO**

248 · · · References and Further Reading

Suarez, R. K. 1998. Oxygen and the upper limits to animal design and performance. *J. Exp. Biol.* 201: 1065–1072. **OX, AE, CO, SY**

Suarez, R. K., J. R. B. Lighton, G. S. Brown, and O. Mathieu-Costello. 1991. Mitochondrial respiration in hummingbird flight muscles. *Proc. Nat. Acad. Sci.* 88: 4870–4873. **MI, CO**

Taylor, C. R. 1994. Relating mechanics and energetics during exercise. In J. H. Jones, ed., *Comparative Vertebrate Exercise Physiology: Unifying Physiological Principles*. Advances in Veterinary Science and Comparative Medicine, vol. 38A. San Diego: Academic Press. Pp. 181–215. **EN, MU**

Taylor, C. R., R. H. Karas, E. R. Weibel, and H. Hoppeler. 1987. Adaptive variation in the mammalian respiratory system in relation to energetic demand. *Resp. Physiol.* 69: 1–127. **OX, AE, SY**

Taylor, C. R., G. M. O. Maloiy, E. R. Weibel, V. A. Langman, J. M. Z. Kamau, M. J. Seeherman, and N. C. Heglund. 1981. Design of the mammalian respiratory system. III. Scaling maximum aerobic capacity to body mass: wild and domestic mammals. *Respir. Physiol.* 44: 25–37. **OX, AE, CO, PH**

Tenney, S. M., and J. E. Remmers. 1963. Comparative quantitative morphology of mammalian lungs: diffusion areas. *Nature* 197: 54–56. **GA, MO, CO**

Turner, D. L., H. Hoppeler, J. Hokanson, and E. R. Weibel. 1995. Cold acclimation and endurance training in guinea pigs: changes in daily and maximal metabolism. *Resp. Physiol.* 101: 183–188. **GA, AE**

Weibel, E. R. 1998. Symmorphosis and optimization of biological design: introduction and questions. In E. R. Weibel, C. R. Taylor, L. Bolis, eds., *Principles of Animal Design: The Optimization and Symmorphosis Debate*. Cambridge: Cambridge University Press. Pp. 1–10. **SY**

Weibel, E. R., H. Claassen, P. Gehr, S. Sehovic, and P. H. Burri. 1980. The respiratory system of the smallest mammal. In K. Schmidt-Nielsen, L. Bolis, C. R. Taylor, eds., *Comparative Physiology: Primitive Mammals*. Cambridge: Cambridge University Press. Pp. 181–191. **GA, CO, MO**

Weibel, E. R., L. B. Marques, M. Constantinopol, F. Doffey, P. Gehr, and C. R. Taylor. 1987. Adaptive variation in the mammalian respiratory system in relation to energetic demand. VI. The pulmonary gas exchanger. *Respir. Physiol.* 69: 81–100. **GA, CO, SY**

Widmer, H. R., H. Hoppeler, E. Nevo, C. R. Taylor, and E. R. Weibel. 1997. Working underground: respiratory adaptations in the blind mole rat. *Proc. Nat. Acad. Sci.* 94: 2062–2067. **GA, PH, MO, CO**

8. Adding Complexity in Form and Function

Keywords: ADipocytes AErobic capacity BLood flow CApillaries COmparative physiology FAtty acids GI–gastro-intestinal system GLucose LIver MItochondria MOrphology MR–metabolic rate MUscle cells OXygen pathway SArcolemma STores, intracellular SUbstrates SYmmorphosis

Further Reading

Diamond, J. M. 1993. Evolutionary physiology. In C. A. R. Boyd and D. Noble, eds. *The Logic of Life: The Challenge of Integrative Physiology.* Oxford: Oxford University Press. Pp. 89–112. **MR, SY**

Diamond, J., and K. Hammond. 1992. The matches, achieved by a natural selection, between biological capacities and their natural loads. *Experientia* 48: 551–557. **GI, SU, SY**

Hammond, K. A., and J. M. Diamond. 1992. An experimental test for a ceiling on sustained metabolic rate in lactating mice. *Physiol. Zool.* 65: 952–977. **SU, GI, SY**

Hofmann, R. R. 1988. Morphophysiological evolutionary adaptations of the ruminant digestive system. In A. Dobson and M. Dobson, eds., *Aspects of Digestive Physiology in Ruminants.* Ithaca, N.Y.: Cornell University Press. Pp. 1–20. **GI, CO**

———1998. How ruminants adapt and optimize their digestive system "blueprint" in response to source shifts. In E. R. Weibel, C. R. Taylor, and L. Bolis, eds., *Principles of Animal Design: The Optimization and Symmorphosis Debate.* Cambridge: Cambridge University Press. Pp. 220–229. **GI, CO**

Hoppeler, H. 1998. The converging pathways for oxygen and substrates in muscle mitochondria. In E. R. Weibel, C. R. Taylor, and L. Bolis, eds., *Principles of Animal Design: The Optimization and Symmorphosis Debate.* Cambridge: Cambridge University Press. Pp. 255–262. **OX, SU, SY**

Hoppeler, H., and E. R. Weibel. 1998. Limits for oxygen and substrate transport in mammals. *J. Exp. Biol.* 201: 1051–1064. **SU, OX, SY**

Hume, I. D. 1989. Optimal digestive strategies in mammalian herbivores. *Physiol. Zool.* 62: 1145–1163. **GI, CO**

Jones, J. H., and S. L. Lindstedt. 1993. Limits to maximal performance. *Ann. Rev. Physiol.* 55: 547–569. **OX, AE, SY**

Pappenheimer, J. R. 1993. On the coupling of membrane digestion with intestinal absorption of sugars and amino acids. *Am. J. Physiol.* 265: G409–G417. **GI, GL**

Peterson, C. C., K. A. Nagy, and J. Diamond. 1990. Sustained metabolic scope. *Proc. Nat. Acad. Sci.* 87: 2324–2328. **CO, MR**

Stevens, C. E., and I. D. Hume. 1995. *Comparative Physiology of the Vertebrate Digestive System.* 2nd ed. Cambridge: Cambridge University Press. **GI, CO**

Weber, J.-M. 1998. Adjusting maximal fuel delivery for differences in demand. In E. R. Weibel, C. R. Taylor, and L. Bolis, eds., *Principles of Animal Design: The Optimization and Symmorphosis Debate.* Cambridge: Cambridge University Press. Pp. 249–254. **SU, AE, PH, SY**

Weibel, E. R., C. R. Taylor, and H. Hoppeler. 1992. Variations in function and design: testing symmorphosis in the respiratory system. *Respir. Physiol.* 87: 325–348. **OX, AE, SY**

Weibel, E. R., C. R. Taylor, J.-M. Weber, R. Vock, T. J. Roberts, and H. Hoppeler.

1996. Design of the oxygen and substrate pathways. I–VII. *J. Exp. Biol.* 199: 1643–1709. **OX, SU, SY**

Wieser, W. 1994. Cost of growth in cells and organisms: general rules and comparative aspects. *Biol. Rev.* 68: 1–33. **MR, CO**

References

Adamson, R. H., and C. C. Michel. 1993. Pathways through the intercellular clefts of frog mesenteric capillaries. *J. Physiol.* 466: 303–327. **CA**

Blouin, A., R. P. Bolender, and E. R. Weibel. 1977. Distribution of organelles and membranes between hepatocytes and nonhepatocytes in the rat liver parenchyma. *J. Cell Biol.* 72: 441–455. **LI**

Crone, C., and D. G. Levitt. 1984. Capillary permeability to small solutes. In E. M. Renkin and C. C. Michel, eds., *Handbook of Physiology*. Sect. 2, vol. IV: *Microcirculation*. Washington, D.C.: American Physiological Society. Pp. 411–466. **CA**

Curry, F. E. 1984. Mechanics and thermodynamics of transcapillary exchange. In E. M. Renkin and C. C. Michel, eds., *Handbook of Physiology*. Sect. 2, vol. IV: *Microcirculation*. Washington, D.C.: American Physiological Society. Pp. 309–374. **CA**

Ferraris, R. P., and J. M. Diamond. 1986. Use of phlorizin binding to demonstrate induction of intestinal glucose transporters. *J. Membrane Biol.* 94: 77–82. **GI, GL**

Folkow, B., and E. Neil. 1971. *Circulation*. New York: Oxford University Press. **BL**

Gollnick, P. D., and B. Saltin. 1988. Fuel for muscular exercise: role of fat. In E. S. Horton and R. L. Terjung, eds., *Exercise, Nutrition and Energy Metabolism*. New York: Macmillan. Pp. 72–87. **MU, FA**

Goodyear, L. J., M. F. Hirshman, R. J. Smith, and E. S. Horton. 1991. Glucose transporter number, activity, and isoform content in plasma membranes of red and white skeletal muscle. *Am. J. Physiol.* 261: E556–E561. **SA, GL**

Hammond, K. A. 1998. The match between load and capacity during lactation: where is the limit to energy expenditure? In E. R. Weibel, C. R. Taylor, and L. Bolis, eds. *Principles of Animal Design: The Optimization and Symmorphosis Debate*. Cambridge: Cambridge University Press. Pp. 205–211. **GI, SU**

Hammond, K. A., M. Konarzewski, R. Torres, and J. M. Diamond. 1994. Metabolic ceilings under a combination of peak energy demands. *Physiol. Zool.* 68: 1479–1506. **MR, SU, AE**

Hammond, K. A., K. C. K. Lloyd, and J. M. Diamond. 1996. Is mammary output capacity limiting to lactational performance. *J. Exp. Biol.* 199: 337–339. **MR, SU**

Hofmann, R. R. 1989. Evolutionary steps of ecophysiological adaptation and diversification of ruminants: a comparative view of their digestive system. *Oecologia* 78: 443–457. **GI, CO**

Hoppeler, H., S. R. Kayar, H. Claassen, E. Uhlmann, and R. H. Karas. 1987. Adaptive variation in the mammalian respiratory system in relation to energetic demand. III. Skeletal muscles: setting the demand for oxygen. *Respir. Physiol.* 69: 27–46. **OX, MU, SY**

Hume, I. D. 1998. Optimization in design of the digestive system. In E. R. Weibel, C. R. Taylor, and L. Bolis, eds., *Principles of Animal Design: The Optimization and Symmorphosis Debate.* Cambridge: Cambridge University Press. Pp. 212–219. **GI, CO**

Jones, J. H., E. K. Birks, and J. R. Pascoe. 1992. Factors limiting aerobic performance. In J.E.P.W. Bicudo, ed., *The Vertebrate Oxygen Transport Cascade: From Atmosphere to Mitochondria.* Boca Raton, Fla.: CRC Press. Pp. 169–178. **AE, OX**

Konarzewski, M., and J. M. Diamond. 1994. Peak sustained metabolic rate and its individual variation in cold-stressed mice. *Physiol. Zool.* 67: 1186–1212. **MR**

Le Maho, Y., and M. Bnouham. 1998. Fuel specialists for endurance. In E. R. Weibel, C. R. Taylor, and L. Bolis, eds., *Principles of Animal Design: The Optimization and Symmorphosis Debate.* Cambridge: Cambridge University Press. Pp. 263–268. **SU, MR**

Madara, J. L. 1998. Regulation of the movement of solutes across tight junctions. *Ann. Rev. Physiol.* 60: 143–159. **GI, GL**

Madara, J. L., and J. R. Pappenheimer. 1987. The structural basis for physiological regulation of paracellular pathways in intestinal epithelia. *J. Membrane Biol.* 100: 149–164. **GI, GL**

Mueckler, M. 1994. Facilitative glucose transporters. *Europ. J. Biochem.* 219: 713–725. **GL, SA**

Pappenheimer, J. R. 1953. Passage of molecules through capillary walls. *Physiol. Rev.* 33: 387–423. **CA**

Pappenheimer, J. R., and K. Z. Reiss. 1987. Contribution of solvent drag through intercellular junctions to absorption of nutrients in the small intestine of the rat. *J. Membrane Biol.* 100: 123–136. **GL, GI**

Roberts, T. J., J.-M. Weber, H. Hoppeler, E. R. Weibel, and C. R. Taylor. 1995. Design of the oxygen and substrate pathways. II. Defining the upper limits of carbohydrate and fat oxidation. *J. Exp. Biol.* 199(8): 1651–1658. **SU, AE, SY**

Taylor, C. R., R. H. Karas, E. R. Weibel, and H. Hoppeler. 1987. Adaptive variation in the mammalian respiratory system in relation to energetic demand. *Resp. Physiol.* 69: 1–127. **OX, AE, SY**

Taylor, C. R., E. R. Weibel, J.-M. Weber, R. Vock, H. Hoppeler, T. J. Roberts, and G. Brichon. 1996. Design of the oxygen and substrate pathways. *J. Exp. Biol.* 199: 1643–1709. **OX, SU, SY**

Turner, D. L., H. Hoppeler, J. Hokanson, and E. R. Weibel. 1995. Cold acclimation and endurance training in guinea pigs: changes in daily and maximal metabolism. *Resp. Physiol.* 101: 183–188. **MR, OX**

Vock, R., E. R. Weibel, H. Hoppeler, G. Ordway, J.-M. Weber, and C. R. Taylor. 1996a. Design of the oxygen and substrate pathways. V. Structural basis of vascular substrate supply to muscle cells. *J. Exp. Biol.* 199(8): 1675–1688. **CA, SU**

Vock, R., H. Hoppeler, H. Claassen, D. X. Y. Wu, R. Billeter, J.-M. Weber, C. R. Taylor, and E. R. Weibel. 1996b. Design of the oxygen and substrate pathways. VI. Structural basis of intracellular substrate supply to mitochondria in muscle cells. *J. Exp. Biol.* 199(8): 1689–1697. **MI, ST, SU**

Wade, O. L., and J. M. Bishop. 1962. *Cardiac Output and Regional Blood Flow.* Oxford: Blackwell Scientific. **BL**

Weber, J.-M., G. Brichon, G. Zwingelstein, G. McClelland, Chr. Saucedo, E. R. Weibel, and C. R. Taylor. 1996b. Design of the oxygen and substrate pathways. IV. Partitioning energy provision from fatty acids. *J. Exp. Biol.* 199(8): 1667–1674. **FA, MR**

Weber, J.-M., T. J. Roberts, R. Vock, E. R. Weibel, and C. R. Taylor, 1996a. Design of the oxygen and substrate pathways. III. Partitioning energy provision from carbohydrates. *J. Exp. Biol.* 199(8): 1659–1666. **GL, MR**

Weibel, E. R., C. R. Taylor, and H. Hoppeler. 1991. The concept of symmorphosis: a testable hypothesis of structure-function relationship. *Proc. Nat. Acad. Sci.* 88: 10357–10361. **SY, OX**

Weibel, E. R., C. R. Taylor, J.-M. Weber, R. Vock, T. J. Roberts, and H. Hoppeler. 1996. Design of the oxygen and substrate pathways. VII. Different structural limits for O_2 and substrate supply to muscle mitochondria. *J. Exp. Biol.* 199(8): 1699–1709. **SU, OX, SY**

Weiner, J. 1993. Physiological limits to sustainable energy budgets in birds and mammals: ecological implications. *Tree* 7: 384–388. **MR**

Wieser, W. 1998. Optimality in complex dynamic systems: constraints, trade-offs, priorities. In E. R. Weibel, C. R. Taylor, and L. Bolis, eds., *Principles of Animal Design: The Optimization and Symmorphosis Debate.* Cambridge: Cambridge University Press. Pp. 230–238. **MR, SY**

9. Symmorphosis in Form and Function

Keywords: ADaptation COmparative physiology EConomy EVolution FRactals INtegration OPtimization OXygen pathway SYmmorphosis

Further Reading

Alexander, R. McN. 1982. *Options for Animals.* London: Arnold. **AD, OP**

Arnold, S. J. 1992. Constraints on phenotypic evolution. *American Naturalist* 140(suppl.): S85–S107. **EV**

Boyd, C. A. R., and D. Noble. 1993. *The Logic of Life: The Challenge of Integrative Physiology.* Oxford: Oxford University Press. **IN**

Diamond, J. M. 1993. Evolutionary physiology. In C. A. R. Boyd and D. Noble, eds., *The Logic of Life: The Challenge of Integrative Physiology.* Oxford: Oxford University Press. Pp. 89–112. **IN, AD**

Dupré, J. 1987. *The Latest on the Best: Essays on Evolution and Optimality.* Cambridge, Mass.: MIT Press. **EV, OP**

Feder, M. E., A. F. Bennett, W. W. Burggren, and R. B. Huey. 1987. *New Directions in Ecological Physiology.* New York: Cambridge University Press. **EV, AD, IN**

Mandelbrot, B. B. 1983. *The Fractal Geometry of Nature.* 2nd ed. San Francisco: Freeman. **FR**

Maynard Smith, J. 1978. Optimization theory in evolution. *Ann. Rev. Ecol. Syst.* 9: 31–56. **EV, OP**

Mayr, E. 1991. *One Long Argument: Charles Darwin and the Genesis of Modern Evolutionary Thought.* Cambridge, Mass.: Harvard University Press. **EV**

Nonnenmacher, T. F., G. A. Losa, and E. R. Weibel. 1994. *Fractals in Biology and Medicine.* Basel: Birkhäuser. **FR**

Parker, G. A., and J. Maynard Smith. 1990. Optimality theory in evolutionary biology. *Nature* 348: 27–33. **OP, EV**

Rose, M. R., and G. V. Lauder. 1996. *Adaptation.* New York and London: Academic Press. **AD**

Schmidt-Nielsen, K. 1984. *Scaling: Why Is Animal Size So Important?* Cambridge: Cambridge University Press. **CO, IN**

Thompson, D'A. W. 1942. *On Growth and Form.* Oxford: Oxford University Press. **AD, OP**

Vogel, S. 1998. Convergence as an analytical tool in evaluating design. In E. R. Weibel, C. R. Taylor, and L. Bolis, eds., *Principles of Animal Design: The Optimization and Symmorphosis Debate.* Cambridge: Cambridge University Press. Pp. 13–20. **AD, OP, CO**

Weibel, E. R., C. R. Taylor, and L. Bolis. 1998. *Principles of Animal Design: The Optimization and Symmorphosis Debate.* Cambridge: Cambridge University Press. **SY, OP, EV**

Weibel, E. R., C. R. Taylor, and H. Hoppeler. 1991. The concept of symmorphosis: a testable hypothesis of structure-function relationship. *Proc. Nat. Acad. Sci.* 88: 10357–10361. **SY, IN, OX**

References

Alexander, R. McN. 1985. The ideal and the feasible: physical constraints on evolution. *Biol. J. Linnean Soc.* 26: 345–358. **EV, AD, OP**

Dudley, R., and C. Gans. 1991. A critique of symmorphosis and optimality models in physiology. *Physiol. Zool.* 64: 627–637. **SY, OP, EV**

Feder, M. E., and W. B. Watt. 1993. Functional biology of adaptation. In R. J. Berry, T. J. Crawford, and G. M. Hewitt, eds., *Genes in Ecology.* Oxford: Blackwell Scientific Publications. Pp. 365–391. **AD**

Garland, T., and R. B. Huey. 1987. Testing symmorphosis: does structure match functional requirements? *Evolution* 41: 1404–1409. **SY**

Garland, T., Jr., and P. A. Carter. 1994. Evolutionary physiology. *Ann. Rev. Physiol.* 56: 579–621. **EV**

Gould, S. J., and R. C. Lewontin. 1979. The spandrels of San Marco and the Panglossian paradigm: a critique of the adaptationist programme. *Proc. Roy. Soc. London* B205: 581–598. **EV**

Hess, W. R. 1914. Das Prinzip des kleinsten Kraftverbrauches im Dienste hämodynamischer Forschung. *Arch. Anat. Physiol.* Physiolog. Abt. **EC, OP**

Hofmann, R. R. 1998. How ruminants adapt and optimize their digestive system "blueprint" in response to resource shifts. In E. R. Weibel, C. R. Taylor, and L. Bolis, eds., *Principles of Animal Design: The Optimization and Symmorphosis Debate.* Cambridge: Cambridge University Press. Pp. 220–229. **AD, OP, EV**

LaBarbera, M. 1990. Principles of design of fluid transport systems in zoology. *Science* 249: 992. **EV, AD**

Maina, J. N. 1998. *The Gas Exchangers: Structure, Function and Evolution of the Respiratory Processes.* Zoophysiology, vol. 37. Berlin and Heidelberg: Springer. 498 pp. **OX, CO, FR, EV, SY**

Maynard Smith, J., R. Burian, S. Kauffman, R. Alberch, J. Campbell, B. Goodwin, R. Lande, D. Raup, and L. Wolpert. 1985. Developmental constraints and evolution. *Quart. Rev. Biol.* 60: 265–287. **EV**

Peitgen, H.-O., and P. H. Richter. 1986. *The Beauty of Fractals.* Berlin-Heidelberg: Springer-Verlag. **FR**

Taylor, C. R., R. H. Karas, E. R. Weibel, and H. Hoppeler. 1987. Adaptive variation in the mammalian respiratory system in relation to energetic demand. *Resp. Physiol.* 69: 1–127. **OX, SY, CO**

Taylor, C. R., and E. R. Weibel. 1981. Design of the mammalian respiratory system. *Resp. Physiol.* 44: 1–164. **OX, SY, CO**

Weibel, E. R. 1984. *The Pathway for Oxygen: Structure and Function in the Mammalian Respiratory System.* Cambridge, Mass.: Harvard University Press. **OX, IN, SY**

——— 1991. Fractal geometry: a design principle for living organisms. *Am. J. Physiol.* 261: L361–L369. **FR**

——— 1994. The significance of fractals for biology and medicine: an introduction summary. In T. Nonnenmacher, G. A. Losa, and E. R. Weibel, eds., *Fractals in Biology and Medicine.* Basel: Birkhäuser. Pp. 2–6. **FR**

Weibel, E. R., C. R. Taylor, and H. Hoppeler. 1992. Variations in function and design: testing symmorphosis in the respiratory system. *Resp. Physiol.* 87: 325–348. **SY, OX**

Wigglesworth, V. B. 1972. *The Principles of Insect Physiology.* 7th ed. London: Chapman & Hall. **CO, EV, OX**

Wigglesworth, V. B., and W. M. Lee. 1982. The supply of oxygen to flight muscle of insects: a theory of tracheole physiology. *Tissue and Cell* 14: 501–518. **OX, CO**

Wilson, E. O. 1992. *The Diversity of Life.* Cambridge, Mass.: Belknap Press of Harvard University Press. **EV**

Credits

Figure 1.3 Reprinted with permission of *Nature* and D. F. Hoyt from D. F. Hoyt and D. R. Taylor, "Gait and Energetics of Locomotion in Horses," 292: 239, copyright 1981 Macmillan Magazines Limited.

Figure 1.6 Reprinted with permission of Cambridge University Press and R. McN. Alexander from E. R. Weibel, C. R. Taylor, and L. Bolis, eds., *Principles of Animal Design: The Optimization and Symmorphosis Debate* (Cambridge, 1998), fig. 2.5, p. 32.

Figures 2.1c, 2.6a, 2.9, 4.1, 4.3, 4.5, 4.6, 4.7, 5.2, 5.3b, and 6.4 From *The Pathway for Oxygen* by E. R. Weibel, copyright © 1984 by the President and Fellows of Harvard College, reprinted by permission of Harvard University Press.

Figure 2.3 Reprinted with permission of the American Physiological Society and P. Cerretelli from R. Margaria, P. Cerretelli, et al., "Kinetics and Mechanism of Oxygen Debt Contraction in Man," *Journal of Applied Physiology* 18 (1963): 371–377, fig. 4.

Figure 2.6 Courtesy K. Schwerzmann.

Figure 2.7 Reprinted with permission of Springer-Verlag from H. Hoppeler, P. Luthi, H. Claassen, E. R. Weibel, and H. Howard, "The Ultrastructure of the Normal Human Skeletal Muscle," *Pfluegers Arch.* 344 (1973), fig. 4, p. 226, copyright 1973 Springer-Verlag.

Figures 3.1, 3.2, 3.6 Reprinted with permission of the Company of Biologists Ltd. from R. Vock et al., "Design of the Oxygen and Substrate Pathways, V and VI," *Journal of Experimental Biology* 199(8)(1996): 1678, 1690–91, figs. 1, 2a–b, 3c.

Figure 3.3 Reprinted with permission of the Company of Biologists Ltd. from T. J. Roberts, J.-M. Weber, H. Hoppeler, E. R. Weibel, and C. R. Taylor, "Design of the Oxygen and Substrate Pathways. II. Defining the Upper Limits of Carbohydrate and Fat Oxidation." *Journal of Experimental Biology* 199(8)(1996): 1651–58.

Figures 3.4, 3.7, 8.1 Reprinted with permission of the Company of Biologists Ltd. from E. R. Weibel et al., "Design of the Oxgen and Substrate Pathways. VII. Different Structural Limits for O_2 and Substrate Supply to Muscle Mitochondria," *Journal of Experimental Biology* 199(13)(1996): 1700–1702.

Figure 4.8 Reprinted from *Respiratory Physiology* 11, A. Geelhaar and E. R. Weibel, "Morphometric Estimation of Pulmonary Diffusion Capacity. III: The Effect of Increased Oxygen Consumption in Japanese Waltzing Mice," copyright 1971, fig. 1, p. 357, with permission from Elsevier Science.

Figure 5.1 From E. R. Weibel and J. Gil, "Structure-Function Relationships at the Alveolar Level," in *Bioengineering Aspects of the Lung,* ed. J. B. West (New York: Marcel Dekker, 1977), fig. 20, by courtesy of Marcel Dekker, Inc.

Figure 5.4a Reprinted with permission of the American Physiological Society from J. Gil et al., "Alveolar Volume-Surface Area Relation," *Journal of Applied Physiology* 47(5)(1979): 990–1007, fig. 6a.

Figures 5.5, 5.9 Reprinted with permission of the McGraw-Hill Companies from E. R. Weibel, "Design and Structure of the Human Lung," in *Pulmonary Diseases and Disorders,* ed. A. P. Fishman (New York: McGraw-Hill, 1980), figs. 15–29, 15–31.

Figure 5.6 Reprinted with permission of the American Physiological Society from H. Bachofen, P. Gehr, and E. R. Weibel, "Alterations of Mechanical Properties and Morphology in Excised Rabbit Lungs Rinsed with a Detergent," *Journal of Applied Physiology* 47 (1979): 1003, fig. 1.

Figure 5.7 Reprinted with permission of the American Physiological Society, T. A. Wilson, and H. Bachofen from T. A. Wilson and H. Bachofen, "A Model for Mechanical Structure of the Alveolar Duct," *Journal of Applied Physiology* 52 (1982): 1064–1070, figs. 1b and c.

Figure 6.2 Reprinted with permission of Lippincott-Raven Publishers and P. H. Burri from P. H. Burri, "Postnatal Development and Growth," in *The Lung: Scientific Foundations,* 2nd ed., ed. R. G. Crystal, J. B. West, E. R. Weibel, and P. J. Barnes (© 1997, by Lippincott-Raven Publishers, Philadelphia), fig. 2.

Figure 6.4 Reprinted with permission of Springer-Verlag from E. R. Weibel, *Morphometry of the Human Lung* (Berlin, 1963), fig. 82, copyright Springer-Verlag.

Figure 6.7 Reprinted with permission of Lippincott-Raven Publishers from E. R. Weibel, "Design of Airways and Blood Vessels Considered as Branching Trees," in R. G. Crystal, J. B. West, E. R. Weibel, and P. J. Barnes, eds., *The Lung: Scientific Foundations,* vol. 1, 2nd ed. (© 1997, by Lippincott-Raven Publishers, Philadelphia), pp. 1061–1071.

Figure 6.8 From B. Mandelbrot, *The Fractal Geometry of Nature,* 2d ed. (San Francisco: W. H. Freeman Co., 1983), pl. 165 (top). Reprinted with permission of the author.

Figure 6.9 Reprinted with permission of the American Physiology Society from E. R. Weibel, "Fractal Geometry," *American Journal of Physiology* 261 (1991): L361–L369, fig. 8.

Figure 7.1 Reprinted with permission from E. R. Weibel, "Oxygen Demand and the Site of Respiratory Structures," in *Evolution of Respiratory Processes,* ed. S. C. Wood and C. Lenfant (New York: Marcel Dekker, 1979), fig. 20, by courtesy of Marcel Dekker, Inc.

Figure 7.2, 7.3 Reprinted from *Respiratory Physiology,* 87, E. R. Weibel, C. R. Taylor, and H. Hoppeler, "Variations in Function and Design," 325–348, figs. 1–2, copyright 1992, with permission from Elsevier Science.

Figure 8.5 Reprinted with permission of Birkhauser Verlag and J. Diamond from J. Diamond and K. Hammond, "The Matches, Achieved by Natural Selection, between Biological Capacities," *Experientia* 48 (1992): 551–557, fig. 2.

Figure 9.1a Reprinted with permission from V. B. Wigglesworth, *Principles of Insect Physiology* (London: Chapman & Hall, 1972), fig. 233, p. 368. Copyright CRC Press, Boca Raton, Florida.

Figure 9.1b Reprinted with permission of Churchill Livingstone from V. B. Wigglesworth and W. M. Lee, "The Supply of Oxygen to the Flight Muscles of Insects," *Tissue and Cell* 14 (1982): 501–508, fig. 15.

Figure 9.4a, b Reprinted with permission of Cambridge University Press and S. Vogel from S. Vogel, "Convergence as an Analytical Tool in Evaluating Design," in *Principles of Animal Design: The Optimization and Symmorphosis Debate*, ed. E. R. Weibel, C. R. Taylor, and L. Bolis (Cambridge, 1998).

Figure 9.4c Courtesy M. LaBarbera.

Figure 9.5 Reprinted with permission of Birkhauser Verlag from E. R. Weibel, "Design of Biological Organisms and Fractal Geometry," in T. Nonnenmacher, G. A. Losa, and E. R. Weibel, eds., *Fractals in Biology and Medicine* (Basel-Boston-Berlin, 1994), pp. 68–85.

Figure 9.6b Reprinted with permission of Springer-Verlag and H. O. Peitgen from H. O. Peitgen and P. H. Richter, *The Beauty of Fractals* (Berlin-Heidelberg, 1986), map 42.

Index

NOTE: Page numbers in *italics* indicate tables and figures.

Acetyl-CoA, 31–33, 61, 65, *182*
Acinus, 144–145, 177
Adaptation (structure/function), 8–11, 151, 211, 218–221. *See also* Variation, adaptive
Adaptive variation. *See* Variation, adaptive
Adipocytes, 65, 196
ADP, 27, 29, *30*, 31–32, *39*, 182
Aerobic capacity (= maximal rate of oxygen consumption, $\dot{V}_{O_2}$max), 22, 34, *34*, 41–59, 81, 151–180, 223; and muscle mitochondria, 35, 40, 45–52, *49*, 161–165, *164, 165*; and muscle capillaries, 35, 49–59, 161, 166–168, *167, 168*; and heart, cardiac output, 161, 167–173, *169, 170*; and pulmonary gas exchanger, 101, 161, 173–178, *175*, 176
Airway diameter, 136–149, *137, 143, 148*, 177
Airways, pulmonary, 81–84, *82, 83, 84*, 132–149, *134–138, 143, 144, 148*, 223, 224, *224*
Allometry. *See* Variation, allometric
Alveolar epithelium, 87, 116, 120, *121*, 125–130, *126, 127*; type I cells, 125, *126, 127*, 129; type II cells, 120, *121*, 125, *127*, 129
Alveolar septum. *See* Septum, alveolar
Alveolar surface area, 84, 90–109, *99, 104*, 110
Alveoli, 82–86, *83, 84, 85*, 133–135, *134*
Anaerobic metabolism. *See* Phosphorylation, anaerobic
Arteries, pulmonary, 81, *82, 85*, 124 (pressure), 132–135, *136*, 143, 149, 223
Athletic performance: human, 45–46, *47*, 101, 168, 197; other species, 48, 76, *105, 105*, 155, 156, 166. *See also* Pronghorn
Athletic/sedentary, *164, 168, 169, 175*. *See also* Variation, adaptive
ATP, 27, 28–40, *30, 38, 39*, 60, 61, 181, *182*

Barriers, lung, 83, 84, *86, 87*, 90, 92–94, 98, 109, 110–130, *128*
Barriers, muscle, 61, 71–74, *72, 74*, 201–203, *202*. *See also* Endothelium, skeletal muscle; Sarcolemma
Barrier thickness (lung), 90–109; harmonic mean, *93, 95, 98*, 174
Basement membrane, lung, 127; fusion of, *126, 127, 128*
β-oxidation of fatty acids, 31–33, 38, 40, 60, 77, *182*
Blood flow, 5, 6, *16, 17, 58*, 124, 139–142, *195, 196*
Blood vessels, design of, *17*, 140–142
Blood viscosity, *58*, 59
Bohr equation, 89, 92
Bohr integration (gradual oxygen loading), 100, 104, 106
Branching of vessels, 17, 132, 135–149, *136, 137, 138, 143, 147*, 224, *224*
Bronchi. *See* Airways, pulmonary

Calorimetry, indirect, 63, 68
Capacity, functional (= maximal functional performance), 19, 22. *See also* Aerobic capacity; Oxygen consumption, maximal rate of
Capillaries, lung, 83, 85–101, *85, 86, 87*, 110, 114, *115*, 177; transit time, *88*, 89, 177; volume, 91, 94–97, *99, 104*; surface area, 93, 95, 97, *99, 104*
Capillaries, skeletal muscle, 30, *30*, 33, *49*, 52–58, *52, 54*, 61–79, *72*, 161; muscle capillary volume and $\dot{V}_{O_2}$max, 35, 52–59, 166–167, *167, 168*; and substrate supply, 65, 71–79, *72*, 195–196, 201–209; and oxygen supply, *72*, 75, 78–79, 207, 222
Carbohydrate pathway, 32, *182*, 184–209
Cardiac output, 5–7, *6*, 124, 167–171, *170, 195, 196*
Carnitine shuttle, 66, 77

· 259 ·

Cell: structure of, 24–26; cytoskeleton, 25; structure of skeletal muscle cell, 26, 27–29. *See also* Muscle *entries*
Circulation, 150, 161, 167–173, *182*, 195, 218, 220
Comparative (physiology): muscle, 47–59, *49*; lung, 101–107, *104, 105*; respiratory system (lung-mitochondria), 154–180
Conductance, lung for oxygen. *See* Diffusing capacity, lung
Contraction: muscle cell, 27–29
Convergence (in evolution), 215
Creatine-phosphate, 29, *30*
Cytochrome oxidase, 37

Diameter: airways, 136–149, *137, 143, 148,* 177; pulmonary blood vessels, 140–143
Dichotomy, 132, *133,* 135–149, *137, 138*
Diffusing capacity, lung (DL): morphometrical, 91–100, *99,* 102–108, *104, 108,* 131; DL_{O_2} = oxygen diffusing capacity (lung), 89–109, *99, 104,* 159; DM_{O_2} = membrane diffusing capacity, 90–94, *98, 99, 104;* DeO_2 = oxygen binding to Hb in erythrocytes, 90–95, *98, 99;* physiological, 100, 102, 107 (dog), *108,* 131
Digestive system. *See* Gastro-intestinal system

Economy: in animal design, 12–21, 58, 78–79, *95,* 109, 111, 149, 151, 195, 211, 221–222; in blood vessel design, 139–142. *See also* Symmorphosis
Edema, lung, 126–129
Emphysema, pulmonary, 116
Endothelium, lung, 87, 116, 125–130
Endothelium, skeletal muscle, 63, 71, 72, 74, 76, 201–203, *202, 204*
Endurance, athlete, 45, 76, 101, 168, 197; exercise (training), 6, 33, *34,* 46. *See also* Pronghorn
Energy: aerobic generation of, 30–35, 60–61 (*see also* Phosphorylation, oxidative); anaerobic generation of, 30–34, 60 (*see also* Glycolysis); expenditure, minimization of, 14; generation in muscle cell, 27, 28–35, *30;* high energy phosphates (*see* ATP; Creatine-phosphate)
Environment, 11, 210
Epithelium: lung, 87, 116, 120, *121,* 125–130. *See also* Alveolar epithelium

Erythrocytes: muscle capillaries, 55, 61, 72, 78–79; lung, 90
Evolution, 8–11, 16, 212–217. *See also* Variation, adaptive
Excess capacity. *See* Redundancy, excess capacity
Exercise: endurance, 33; oxygen needs in, 33–34, *34,* 81, 99, 150, 153, *153, 156,* 163; substrate supply in, 66–79, *69,* 81, 195–209, *201;* pulmonary gas exchange in, 81, *88,* 89; blood flow partitioning in, 124, 195, *196*

Fat (adipose) tissue, 65, 191, 196
Fatty acids, 30–33, 35–40, 60–79; pathway, 32, *65, 182,* 184–209; fatty acid binding protein, 63
Fiber system, lung. *See* Lung fiber system
Fick's law of diffusion, 89, 168
Fluid (edema), 125–129
Flux density. *See* Glucose, flux densities; Substrate(s)
Form, functional importance of, 3
Fractals, 145–149, *147, 148,* 223–228, *229;* Koch tree, 146–149, *147,* 227
Fuel supply. *See* Substrate supply
Functional vs. design parameters (respiratory system), 160–180

Gait, maximal stride frequency, 12, *13,* 171
Gas exchange(r), pulmonary, 80–109, *88,* 110–130, 132, *134–135,* 161, 173–178, *216–217,* 219–223
Gastro-intestinal system, *182,* 185–194, 216; gut, resorptive surface, 186–194, *187, 188–189;* mammals, general, 185–188; ruminants, 188–191; other herbivores, 191
Genome, 9, 14, 24, 210
Gills, 213, *216*
Glucagon, 195
Gluconeogenesis (liver), 191, 206
Glucose, 30, 31–33, 37–40, 60–79, *182, 204;* pathway, 31–33, *65, 76, 182,* 184–209; transporters (GLUT-1, GLUT-4), 63, 71–73, *182,* 192, 203, 205, 218; flux densities, capillary–muscle cell, 71–73, *201, 203, 204;* intracellular, *182, 201,* 204–206; uptake, 185–198, *194;* circulating, vascular, 199–204, *201*
Glycogen, 63, *64, 75, 182,* 195; stores, general and liver, 63, *182,* 184, 188, *190,*

195–209; stores, intramyocellular, 62–66, *64, 75–78, 182,* 197–209, *201,* 219; repletion, 205
Glycolysis, 30–34, *34,* 60, 198, *199*

Heart, 5, 150, 167–173; rate, 6–7, *6,* 167–173, 218, 220; size, *6, 7,* 167–172, 220
Hematocrit, *49,* 56–59, *58,* 95, 103, *104,* 160, 166, *168,* 169, *169,* 172, 174, 219–222
Hemoglobin, 53, 88–95, 160, 218
Hierarchy (of structural elements), 12, 210
Horse, racehorse, 20, *20,* 166
Hummingbird, 45, 51, 165 (mitochondria), 218 (heart rate)
Hypoxia: high altitude, 106, 179; tolerance, 107

Insulin, 195, 204, 205
Integration (of function), 11–12, 151, 211, 219

Kleiber's equation, curve, 155, *156*
Krebs (tricarboxylic acid) cycle, 31–33, 37–40, 60, 65, *182*
Krogh cylinder, 53–55, *54*
Krogh's permeation coefficient, 90, 93, 99

Lactic acid, lactate, 33, *34,* 181, *199,* 206, 207
Lipid droplets, intramyocellular, 62–66, *64, 75–78, 182,* 191, 196–198, 208–209, 219
Lipids. *See* Triglyceride stores
Liver, 188, *190,* 196–206
Locomotor system, 11, 14, 210, 218
Lung, 80–149, 219; tissue barrier, 83–84, *86, 87,* 90–109, 110–130, *128* (*see also* Barrier thickness); capillaries, *86, 87,* 91–97, *99, 104,* 114, *115,* 217 (*see also* Capillaries, lung); endothelium, *87,* 116, 125–130; epithelium, *87,* 116, 120, *121,* 125–130 (*see also* Alveolar epithelium); oxygen diffusing capacity $D_{L_{O_2}}$, 89–109, *99, 104,* 159 (*see also* Diffusing capacity, lung); volume, 95–98, *99, 104,* 106; surface tension, 111, 116–124; tissue tension, 113, 116, 117, 123, 124; pressure-volume curve, 120, 122, *122;* development, 120, 132–136, *133, 134–135,* 222. *See also* Pulmonary
Lung fiber system, 111–117, *112, 114,* 123–124; axial, 112, *112, 114;* axial, rings around alveolar mouths, 113, *114,* 124; peripheral, *112,* 113, *114;* septal, *112,* 113, *114, 115,* 123

Malleability, 10, 180, 192. *See also* Variation, induced
Mechanics: lung, 110–124, *112, 114, 115, 118–119, 121, 122, 123*
Metabolic rate, 155. *See also* $\dot{V}_{O_2}$max
Metabolism: oxidative in muscle, 24–59, 60–79, 181–206. *See also* Phosphorylation, oxidative
Mice: Japanese Waltzing, 101, *102, 103,* 173
Mitochondria, 21, 25, *26–27,* 29–57, *36–37,* 60–79; function, 31–33, 37–40, *39, 182* (*see also* Phosphorylation, oxidative); membrane, inner (cristae), 31–33, 37–46, *38, 39,* 50–52, 57 (*see also* Respiratory chain); membrane, outer, 35, *38, 39;* structure, 35, *36–37, 38,* 47; volume, 42–56, *47, 49,* 161–165, 219; volume and $\dot{V}_{O_2}$max, 35, 40, 45–52, *47, 49,* 162–165, *164, 165;* intermembrane space, 37, *38, 39;* matrix, 37–44 (*see also* Krebs cycle; β-Oxidation); adaptive variation, 48, *49;* heart, 163
Mole rat *(Spalax ehrenbergi),* 15, *15,* 106, 180, 221
Morphogenesis, 1, 3, 7, 18, 222; lung, 132–136
Morphometry, 22, 23, 157; skeletal muscle, 29, 40–44, 45, *49,* 202; lung, 91–100, *99,* 102, *104,* 107, 174
Murray's law (branched vessels), 138–145, *143,* 148, 224, *224–225*
Muscle, skeletal, 26–59, *26–27,* 30, *34, 36, 38, 39, 47, 52, 53, 54,* 60–79, *202*
Muscle cells: myofilaments (actomyosin), 25–29, *27, 30, 182;* myofibrils, *26, 27;* myoglobin, 52
Myelin, tubular, 120

Nautilus (shell), 227, *228*
Network model, *182*
Nutrients: digestion and uptake, 184–194, 218

Optimization (of design), 16, 19, 95, 142, 212. *See also* Symmorphosis
β-Oxidation of fatty acids, 31–33, 38, 40, 60, 77, *182*

Oxidative phosphorylation. *See* Phosphorylation, oxidative
Oxygen binding to hemoglobin, 88, 94
Oxygen consumption ($\dot{V}_{O_2}$), 5, *6*, 29, *34*, 62–63, *153, 156*; rest–exercise, *34*, 81, 99, 153, *153*, 155, *156*, 163; standard metabolic rate, 155
Oxygen consumption, maximal rate of ($\dot{V}_{O_2}$max), 22, 34, *34*, 41–59, 81, 151–180, 223; and muscle mitochondria, 35, 40, 45–52, *49*, 161–165, *164, 165*; and muscle capillaries, 35, 49–59, 161, 166–168, *167, 168*; and heart, cardiac output, 161, 167–173, *169, 170*; and pulmonary gas exchanger, 101, 161, 173–178, *175, 176*. *See also* $\dot{V}_{O_2}$max
Oxygen flow rates, 158, *159*
Oxygen partial pressures, 86–89, *88, 159, 159*, 161, 177
Oxygen pathway (lung-mitochondria), 150–180, *159*, 181–184, *182*, 206–209, *221*
Oxygen supply: to muscle mitochondria, 5, *6*, 52–59, 61, 75, 78, 181–184, *182*, 206–209; and muscle capillaries, 75, 78–79
Oxygen transport in blood, 160
Oxygen uptake: lung, 80–101; insects (tracheoles), 213, *214*; fish (gills), 213, *216*

Pathway for oxygen (lung-mitochondria), 150–180, *159*, 181–184, *182*, 206–209, *221*
Pathway for oxygen and substrates, 65, *65*, 181–209, *182*
Pathway for substrates, 65–66, *65*; direct, 76 (glucose), *182*, 199–204; indirect, 76 (glucose), *182*, 204–206. *See also* Substrate(s)
Perfusion, lung, 124, 131–149, 219, 223
Permeation coefficient for oxygen, 90, 93, 99
Phospholipid monolayer (alveoli), 119
Phosphorylation: oxidative, 25, 29–33, 40, 60–61, 198, *199*, 200, 218, 220; anaerobic, 30–34, *34*, 60, 198, *199*
Plasma, blood, 61–66, 69–74, *72*, 90
Pneumonectomy, 107–109, *108*
Poiseuille's law (blood flow), 17, 139, 140
Pressure-volume curve, lung, 120, 122, *122*
Pronghorn, 1–4, *2*, 14, 20, 22, 48, *49*, 55, 105–107, *104*, 155, *156*, 179, 221
Proteins, 60, 185, 186

Pulmonary: veins, 81–83, *82*; arteries, 81–83, *82, 85*, 124 (pressure), 132–135, *136*, 143, 149, 223; airways, 81–84, *82, 83, 84*, 132–149, *134–138, 143, 144, 148*, 223, 224, *224*. *See also* Diameter

Red blood cells. *See* Erythrocytes
Redundancy, excess capacity, 101, 109, 110, 131, 179, 180, 223
Respiratory chain, 31–33, 37–40, 50, 57, 159, 218
Respiratory distress syndrome: newborn, premature babies, 120, 129; adult, 129
Respiratory exchange ratio (RER), 62, 68, 81
Respiratory system, 150–180, 213–215, 218–224; model of, 158–162, *159*

Safety factor, 20, 125, 131, 223. *See also* Redundancy, excess capacity
Sarcolemma, 27; as diffusion barrier, 63, 71–75, *75, 76*, 78, *182*, 201–204, *202, 204*, 208
Self-similarity, 145–149, 225. *See also* Fractals
Septum, alveolar, 125–130, *126, 127*
Shock lung, 129
Shrew, Etruscan *(Suncus etruscus)*, 152, *152*, 164, 218 (heart rate)
Size, animal, 8. *See also* Variation, allometric
Stereological methods, 42–44 (muscle), 95, 96–98 (lung), 158
Stores, intracellular. *See* Substrate(s), stores
Substrate(s): flux through barriers (glucose), 71–73, 200–205, *204*, 208; metabolism in mitochondria, 31–33; partitioning, *182*, 207–209; partitioning, carbohydrate vs. fat sources, 62–63, 65, 68–71, *69*; partitioning, intracellular vs. intravascular sources, 62–63, 65, *65*, 69–79; stores, general (liver, adipocytes), 63, 65, *182*, 184 (*see also* Glycogen; Triglyceride stores); stores, intramyocellular, 62–66, *64*, 71, 75–78, *182* (*see also* Glycogen; Triglyceride stores); supply capillary, vascular, 62–63, 69–79, *70*; supply to mitochondria, 60–79, *65, 69*, 181–209, *182, 201*; uptake, 185–199
Surface tension, surface forces, 111, 116–124

Index · · · 263

Surfactant, 117–124, *118–119, 121,* 129, 213
Symmorphosis, 18–22, 149, 151–180, 197–209, 210–229; defined, 18–22; oxygen pathway in muscle, 51, 52, 56–59, 78–79

Tissue tension, lung, 113, 116, 117, 123–124
Triglyceride stores, *182,* 184, 196; general (adipocytes and liver), 196–198; intramyocellular (lipid droplets), 62–66, *64,* 75–78, *182,* 191, 196–198, 208–209, 219

Variation (of form and function): adaptive (evolution) (athletic/sedentary), 8, 10, 22, 48, *49,* 55, *104,* 151, 156, *156,* 162–175, *164, 165, 167, 168, 169, 170, 175, 176,* 220; allometric (size), 8, 10, 22, 152, *155, 156,* 157, 162–175, *165, 167, 170, 176,* 220; induced (malleability), 10
Vascular system, 16, 139
Ventilation, lung, 131–149, 219, 223
Viscosity, blood, *58, 59*
V_{O_2}max (maximal rate of oxygen consumption), 22, 34, *34,* 41–59, 81, 151–180, 223; and muscle mitochondria, 35, 40, 45–52, *49,* 161–165, *164, 165;* and muscle capillaries, 35, 49–59, 161, 166–168, *167, 168;* and heart, cardiac output, 161, 167–173, *169, 170;* and pulmonary gas exchanger, 101, 161, 173–178, *175, 176;* per unit body mass, *48, 49, 155;* per unit mitochondrial volume, 49–51, *49, 162*

GETTYSBURG COLLEGE

3326800 0362134 7